STUDENT'S STUDY GUIDE

BEGINNING ALGEBRA

STUDENT'S STUDY GUIDE

BEGINNING ALGEBRA

SIXTH EDITION

LIAL • MILLER • HORNSBY

Prepared with the assistance of
MARJORIE SEACHRIST

HarperCollins*Publishers*

Cover photo: Comstock/George Gerster

Flushing salts from contoured fields in California's Coachella Valley, farmers flood fallow land with from one to five acre-feet of water, sometimes more than they use to irrigate crops. Soluble salts leach through the soil into buried drains. Caused mainly by use of saline Colorado River water for irrigation, salt buildup can significantly reduce soil productivity.

Student's Study Guide for BEGINNING ALGEBRA, sixth edition.

Copyright © 1992 by HarperCollins Publishers, Inc.

All rights reserved. Printed in the United States of America. No part of this book may be used or reproduced in any manner whatsoever without written permission, except in the case of brief quotations embodied in critical articles and reviews. For information address HarperCollins Publishers Inc., 10 East 53rd Street, New York, NY 10022.

ISBN 0-673-46461-X

92 93 94   9 8 7 6 5 4 3 2

PREFACE

This book is designed to be used along with Beginning Algebra, sixth edition, by Margaret L. Lial, Charles D. Miller, and E. John Hornsby, Jr.

In this book each of the objectives in the textbook is illustrated by a solved example or explained using a modified form of programmed instruction. This book should be used in addition to the instruction provided by your instructor. If you are having difficulty with an objective, find the objective in the Student's Study Guide and carefully read and complete the appropriate section.

Also, in this book you will find a short list of suggestions that may help you to become more successful in your study of mathematics. A careful reading should prove to be a valuable experience.

The following people have made valuable contributions to the production of this Student's Study Guide: Marjorie Seachrist, editor; Judy Martinez, typist; Therese Brown and Charles Sullivan, artists; and Eric Seachrist, proofreader.

We also want to thank Tommy Thompson of Seminole Community College for his suggestions for the essay, "To the Student: Success in Algebra" that follows this preface.

TO THE STUDENT: SUCCESS IN ALGEBRA

The main reason students have difficulty with mathematics is that they don't know how to study it. Studying mathematics *is* different from studying subjects like English or history. The key to success is regular practice.

This should not be surprising. After all, can you learn to play the piano or to ski well without a lot of regular practice? The same thing is true for learning mathematics. Working problems nearly every day is the key to becoming successful. Here is a list of things you can do to help you succeed in studying algebra.

1. *Attend class regularly.* Pay attention in class to what your instructor says and does, and make careful notes. In particular, note the problems the instructor works on the board and copy the complete solutions. Keep these notes separate from your homework to avoid confusion when you read them over later.

2. Don't hesitate to ask questions in class. It is not a sign of weakness, but of strength. There are always other students with the same question who are too shy to ask.

3. *Read your text carefully.* Many students read only enough to get by, usually only the examples. Reading the complete section will help you to be successful with the homework problems. Most exercises are keyed to specific examples or objectives that will explain the procedures for working them.

4. Before you start on your homework assignment, rework the problems the instructor worked in class. This will reinforce what you have learned. Many students say, "I understand it perfectly when you do it, but I get stuck when I try to work the problem myself."

5. Do your homework assignment only *after* reading the text and reviewing your notes from class. Check your work with the answers in the back of the book. If you get a problem wrong and are unable to see why, mark that problem and ask your instructor about it. Then practice working additional problems of the same type to reinforce what you have learned.

6. Work as neatly as you can. Write your symbols clearly, and make sure the problems are clearly separated from each other. Working neatly will help you to think clearly and also make it easier to review the homework before a test.

7. After you have completed a homework assignment, look over the text again. Try to decide what the main ideas are in the lesson. Often they are clearly highlighted or boxed in the text.

8. Use the chapter test at the end of each chapter as a practice test. Work through the problems under test conditions, without referring to the text or the answers until you are finished. You may want to time yourself to see how long it takes you. When you have finished, check your answers against those in the back of the book and study those problems that you missed. Answers are referenced to the appropriate sections of the text.

9. Keep any quizzes and tests that are returned to you and use them when you study for future tests and the final exam. These quizzes and tests indicate what your instructor considers most important. Be sure to correct any problems on these tests that you missed, so you will have the corrected work to study.

10. Don't worry if you do not understand a new topic right away. As you read more about it and work through the problems, you will gain understanding. Each time you look back at a topic you will understand it a little better. No one understands each topic completely right from the start.

CONTENTS

1 THE REAL NUMBER SYSTEM

1.1 Fractions — 1

1.2 Exponents, Order of Operations, and Inequality — 7

1.3 Variables, Expressions, and Equations — 14

1.4 Real Numbers and the Number Line — 19

1.5 Addition of Real Numbers — 24

1.6 Subtraction of Real Numbers — 28

1.7 Multiplication of Real Numbers — 32

1.8 Division of Real Numbers — 36

1.9 Properties of Addition and Multiplication — 39

Chapter 1 Test — 46

2 SOLVING EQUATIONS AND INEQUALITIES

2.1 Simplifying Expressions — 49

2.2 The Addition and Multiplication Properties of Equality — 52

2.3 More on Solving Linear Equations — 59

2.4 An Introduction to Applications of Linear Equations — 66

2.5 Formulas and Applications from Geometry — 71

2.6 Ratios and Proportions — 76

2.7 Applications of Percent: Mixture, Interest, and Money — 80

2.8 More About Problem Solving — 86

2.9 The Addition and Multiplication Properties of Inequality — 89

Chapter 2 Test — 98

3 POLYNOMIALS AND EXPONENTS

3.1	Polynomials	101
3.2	Exponents	108
3.3	Multiplication of Polynomials	112
3.4	Special Products	118
3.5	The Quotient Rule and Integer Exponents	120
3.6	The Quotient of a Polynomial and Monomial	124
3.7	The Quotient of Two Polynomials	125
3.8	An Application of Exponents: Scientific Notation	128
Chapter 3 Test		131

4 FACTORING AND APPLICATIONS

4.1	Factors; The Greatest Common Factor	133
4.2	Factoring Trinomials	136
4.3	More on Factoring Trinomials	139
4.4	Special Factorizations	143
4.5	Solving Quadratic Equations by Factoring	147
4.6	Applications of Quadratic Equations	150
4.7	Solving Quadratic Inequalities	154
Chapter 4 Test		158

5 RATIONAL EXPRESSIONS

5.1	The Fundamental Property of Rational Expressions	161
5.2	Multiplication and Division of Rational Expressions	165
5.3	The Least Common Denominator	169
5.4	Addition and Subtraction of Rational Expressions	172
5.5	Complex Fractions	174
5.6	Equations Involving Rational Expressions	177
5.7	Applications of Rational Expressions	182
	Chapter 5 Test	190

6 GRAPHING LINEAR EQUATIONS

6.1	Linear Equations in Two Variables	193
6.2	Graphing Linear Equations in Two Variables	200
6.3	The Slope of a Line	206
6.4	Equations of a Line	208
6.5	Graphing Linear Inequalities in Two Variables	212
6.6	Functions	216
	Chapter 6 Test	222

7 LINEAR SYSTEMS

7.1	Solving Systems of Linear Equations by Graphing	225
7.2	Solving Systems of Linear Equations by Addition	232
7.3	Solving Systems of Linear Equations by Substitution	240
7.4	Applications of Linear Systems	245
7.5	Solving Systems of Linear Inequalities	250
	Chapter 7 Test	252

8 ROOTS AND RADICALS

8.1	Finding Roots	255
8.2	Multiplication and Division of Radicals	259
8.3	Addition and Subtraction of Radicals	262
8.4	Rationalizing the Denominator	264
8.5	Simplifying Radical Expressions	267
8.6	Equations with Radicals	271
8.7	Fractional Exponents	275
Chapter 8 Test		278

9 QUADRATIC EQUATIONS

9.1	Solving Quadratic Equations by the Square Root Property	280
9.2	Solving Quadratic Equations by Completing the Square	283
9.3	Solving Quadratic Equations by the Quadratic Formula	287
9.4	Complex Numbers	294
9.5	Graphing Quadratic Equations in Two Variables	301
Chapter 9 Test		308

ANSWERS TO CHAPTER TESTS   313

CHAPTER 1 THE REAL NUMBER SYSTEM

1.1 Fractions

1. Learn the definition of *factor*. (See Frames 1–5 below.)
2. Write fractions in lowest terms. (Frames 6–8)
3. Multiply and divide fractions. (Frames 9–15)
4. Add and subtract fractions. (Frames 16–24)
5. Solve problems that involve operations with fractions. (Frames 25–26)

---

1. The answer to a multiplication problem is called the _____.	product
2. The numbers that are multiplied to get a product are called _____. A natural number, except 1, that has only itself and 1 as natural number factors is a _____ number.	factors prime
3. Write 36 as a product of prime factors. $\qquad\qquad$ 36 = _____	$2 \cdot 2 \cdot 3 \cdot 3$
4. Write 77 as a product of prime factors. $\qquad\qquad$ 77 = _____	$7 \cdot 11$
5. Write 41 as a product of prime factors. $\qquad\qquad$ 41 = _____	41
6. It is useful to express fractions in _____ terms, in which the numerator and denominator have no _____ in common other than _____. For example, 2/3 (*is/is not*) expressed in lowest terms. Is 15/20 in lowest terms? (*yes/no*)	lowest factors; 1 is no
7. To write 15/20 in lowest terms, find the greatest common factor of _____ and _____. The number to use here is _____. Divide into the numerator and denominator.	15; 20 5

Chapter 1 The Real Number System

$$\frac{15}{20} = \frac{3 \cdot ()}{2 \cdot 2 \cdot ()} = \underline{}$$ $\frac{3 \cdot (5)}{2 \cdot 2 \cdot (5)}; \frac{3}{4}$

The dot tells us to _____ the numbers.   multiply

8. Write each fraction number in lowest terms.

$\frac{6}{8} = $ _____ $\frac{3}{4}$

$\frac{12}{18} = $ _____ $\frac{2}{3}$

$\frac{15}{24} = $ _____ $\frac{5}{8}$

$\frac{21}{56} = $ _____ $\frac{3}{8}$

$\frac{33}{55} = $ _____ $\frac{3}{5}$

9. To multiply two fractions, first multiply the _____ and then multiply the _____. Then write the answer in lowest terms, if necessary. For example, multiply 2/3 and 5/7.

numerators
denominators

$$\frac{2}{3} \cdot \frac{5}{7} = \frac{()()}{()()} = \underline{}$$ $\frac{(2)(5)}{(3)(7)}; \frac{10}{21}$

Here the product is in _____ terms.   lowest

10. Multiply fractions. Write all products in lowest terms.

$\frac{9}{10} \cdot \frac{5}{8} = $ _____ $\frac{9}{16}$

$\frac{3}{8} \cdot \frac{10}{18} = $ _____ $\frac{1}{3}$

$\frac{21}{30} \cdot \frac{5}{7} = $ _____ $\frac{1}{2}$

$\frac{2}{5} \cdot \frac{5}{9} \cdot \frac{18}{30} = $ _____ $\frac{2}{9}$

1.1 Fractions

11. To divide two fractions, _____ the first fraction times the _____ of the second fraction. For example, divide 3/8 by 3/4.

    $$\frac{3}{8} \div \frac{3}{4} = \frac{3}{4} \cdot \underline{} = \underline{}$$

    multiply
    reciprocal

    $\frac{3}{8} \cdot \frac{4}{3}; \frac{1}{2}$

12. Divide fractions. Write all quotients in in lowest terms.

    $\frac{5}{9} \div \frac{10}{3} = \underline{}$ $\frac{1}{6}$

    $\frac{2}{5} \div \frac{6}{10} = \underline{}$ $\frac{2}{3}$

    $\frac{21}{30} \div \frac{14}{20} = \underline{}$ 1

    $\frac{12}{15} \div \frac{30}{20} = \underline{}$ $\frac{8}{15}$

13. To multiply or divide in a series of calculations, start at the _____, and multiply or divide, as indicated, from _____ to _____. For example, to work $3 \cdot 8 \div 12 \div 3 \cdot 9$, first work _____.

    left
    left; right

    $3 \cdot 8$

    $3 \cdot 8 \div 12 \div 3 \cdot 9 = () \div 12 \div 3 \cdot 9$ 24

    Now work _____. $24 \div 12$

    $= () \div 3 \cdot 9$ 2

    $2 \div 3 = \underline{}$, so we have $\frac{2}{3}$

    $= () \cdot 9$ $\frac{2}{3}$

    $= \underline{}$.   6

14. Calculate $\frac{3}{8} \div \frac{27}{12} \cdot \frac{6}{7} \div \frac{15}{14}$. Start at the left:

    $\frac{3}{8} \div \frac{27}{12} = \underline{}$. Now work $\frac{1}{6} \cdot () = \underline{}$. $\frac{1}{6}; \frac{6}{7}; \frac{1}{7}$

    Finally, $\frac{1}{7} \div \frac{15}{14} = \underline{}$. $\frac{2}{15}$

15. Calculate $\frac{5}{9} \div \frac{8}{12} \div \frac{15}{9} \div \frac{21}{10}$.

    Answer: _____ $\frac{5}{21}$

Chapter 1 The Real Number System

16. To add two fractions having the same _____ just add the _____ and keep the _____. That is,

    $\dfrac{a}{b} + \dfrac{c}{b} =$ _____ .

    For example, $\dfrac{2}{9} + \dfrac{4}{9} = \dfrac{}{9} = \dfrac{}{9}$, which reduces to lowest terms as _____ .

    denominator
    numerators
    denominator

    $\dfrac{a+c}{b}$

    $\dfrac{2+4}{9}$; $\dfrac{6}{9}$

    $\dfrac{2}{3}$

17. Add the fractions. Write all answers in lowest terms.

    $\dfrac{2}{5} + \dfrac{1}{5} =$ _____ $\dfrac{3}{5}$

    $\dfrac{11}{15} + \dfrac{1}{15} =$ _____ $\dfrac{4}{5}$

    $\dfrac{7}{18} + \dfrac{2}{18} =$ _____ $\dfrac{1}{2}$

    $\dfrac{5}{21} + \dfrac{9}{21} =$ _____ $\dfrac{2}{3}$

18. In order to add two fractions that do not have the same _____, it is necessary to find a _____ denominator. For example, a common denominator for 3/4 and 7/10 is _____. There are many other common denominators that could be used, but 20 is the _____ common denominator, and involves less work than any other common denominator. To add 3/4 and 7/10, we must rewrite 3/4 and 7/10 as fractions having a denominator of _____. Do this as follows.

    $\dfrac{3}{4} = \dfrac{?}{20}$

    Divide 4 into _____, getting an answer of _____. Now multiply 5 and _____, getting _____. Therefore,

    $\dfrac{3}{4} = \dfrac{}{20}$. Also, $\dfrac{7}{10} = \dfrac{}{20}$.

    denominator
    common
    20

    least

    20

    20; 5
    3; 15

    $\dfrac{15}{20}$; $\dfrac{14}{20}$

1.1 Fractions 5

19. Write each fraction with the indicated denominator.

$$\frac{9}{10} = \frac{}{50}$$

$\frac{45}{50}$

$$\frac{3}{7} = \frac{}{42}$$

$\frac{18}{42}$

$$\frac{4}{11} = \frac{}{77}$$

$\frac{28}{77}$

20. To add fractions with different denominators, first find a _____ denominator. Use the results of Frame 18 above to add 3/4 and 7/10. (The common denominator was _____.)

common

20

$$\frac{3}{4} + \frac{7}{10} = \frac{}{20} + \frac{}{20} = \frac{}{20}$$

$\frac{15}{20}$; $\frac{14}{20}$; $\frac{29}{20}$

21. Add the fractions. Write all answers in lowest terms. Add from left to right in the last two problems.

$$\frac{2}{3} + \frac{5}{6} = \frac{}{6} + \frac{5}{6} = \frac{}{6} = \frac{}{2}$$

$\frac{4}{6}$; $\frac{9}{6}$; $\frac{3}{2}$

$$\frac{5}{12} + \frac{1}{4} = \frac{5}{12} + \frac{}{12} = \frac{}{12} = \frac{}{3}$$

$\frac{3}{12}$; $\frac{8}{12}$; $\frac{2}{3}$

$$\frac{5}{9} + \frac{5}{6} = \underline{}$$

$\frac{25}{18}$

$$\frac{2}{15} + \frac{7}{20} = \underline{}$$

$\frac{29}{60}$

$$\frac{1}{2} + \frac{2}{3} + \frac{3}{8} = \underline{}$$

$\frac{37}{24}$

$$\frac{3}{4} + \frac{5}{12} + \frac{5}{9} = \underline{}$$

$\frac{31}{18}$

22. There are two methods for adding mixed numbers such as

$$4\frac{2}{3} + 3\frac{1}{2}.$$

Chapter 1 The Real Number System

Method 1

Rewrite both numbers.

$$4\frac{2}{3} = \frac{4}{1} + \frac{2}{3} = \frac{}{3} + \frac{2}{3} = \frac{}{3} \qquad \frac{12}{3} + \frac{2}{3};\ \frac{14}{3}$$

$$3\frac{1}{2} = \frac{3}{1} + \frac{1}{2} = \frac{}{2} + \frac{1}{2} = \frac{}{2} \qquad \frac{6}{2} + \frac{1}{2};\ \frac{7}{2}$$

Now add.

$$4\frac{2}{3} + 3\frac{1}{2} = \frac{14}{3} + \frac{7}{2} = \frac{}{6} + \frac{}{6} = \frac{}{6} \qquad \frac{28}{6} + \frac{21}{6};\ \frac{49}{6}$$

Method 2

Write $4\frac{2}{3}$ as $4\frac{}{6}$ and $3\frac{1}{2}$ as $3\frac{}{6}$. $\qquad 4\frac{4}{6};\ 3\frac{3}{6}$

Then add vertically.

$$\begin{array}{r} 4\frac{2}{3} \\ +\ 3\frac{1}{2} \\ \hline \end{array} \longrightarrow \begin{array}{r} 4\frac{}{6} \\ +\ 3\frac{}{6} \\ \hline \underline{} = \underline{} \end{array}$$

$4\frac{4}{6}$

$3\frac{3}{6}$

$7\frac{7}{6};\ 8\frac{1}{6},\ \text{or}\ \frac{49}{6}$

23. To subtract two rational numbers having the same denominator, _____ the numerators and keep the denominator. For example,   subtract

$$\frac{5}{8} - \frac{3}{8} = \frac{}{8} = \frac{}{8} = \frac{}{4}. \qquad \frac{5-3}{8};\ \frac{2}{8};\ \frac{1}{4}$$

24. Subtract the rational numbers. Write all answers in lowest terms. Subtract from left to right in the last two problems.

$$\frac{5}{12} - \frac{1}{12} = \underline{} \qquad \frac{1}{3}$$

$$\frac{6}{11} - \frac{2}{11} = \underline{} \qquad \frac{4}{11}$$

$$\frac{5}{8} - \frac{1}{3} = \frac{}{24} - \frac{}{24} = \frac{}{24} \qquad \frac{15}{24} - \frac{8}{24};\ \frac{7}{24}$$

$$\frac{9}{10} - \frac{1}{6} = \frac{}{30} - \frac{}{30} = \frac{}{30} = \frac{}{15} \qquad \frac{27}{30} - \frac{5}{30};\ \frac{22}{30};\ \frac{11}{15}$$

1.2 Exponents, Order of Operations, and Inequality

$$\frac{3}{4} - \frac{1}{12} - \frac{1}{3} = \frac{}{12} - \frac{}{12} - \frac{}{12} = \frac{}{12}$$

$$= \frac{}{3}$$

$\frac{9}{12} - \frac{1}{12} - \frac{4}{12}; \frac{4}{12}$

$\frac{1}{3}$

$$\frac{3}{2} - \frac{5}{6} - \frac{1}{4} = \frac{}{24} - \frac{}{24} - \frac{}{24} = \frac{}{24}$$

$$= \frac{}{12}$$

$\frac{36}{24} - \frac{20}{24} - \frac{6}{24}; \frac{10}{24}$

$\frac{5}{12}$

25. Carl types 2/5 of a page in a minute. How many minutes does it take him to type $10\frac{1}{2}$ pages?

    Answer: _____

    $26\frac{1}{4}$ minutes

26. Hannah collected $10\frac{3}{4}$ pounds of newspapers for recycling on Monday, $12\frac{1}{3}$ pounds on Tuesday, and $9\frac{1}{2}$ pounds on Wednesday. How many pounds of newspaper did Hannah collect on the three days?

    Answer: _____

    $32\frac{7}{12}$ pounds

1.2 Exponents, Order of Operations, and Inequality

[1] Use exponents. (See Frames 1-12 below.)

[2] Use the order of operations. (Frames 13-26)

[3] Use more than one grouping symbol. (Frames 27-32)

[4] Know the meanings of $\neq$, $<$, $>$, $\leq$, and $\geq$. (Frames 33-40)

[5] Translate word statements to symbols. (Frames 41-45)

[6] Write statements that change the direction of inequality symbols. (Frames 46-51)

---

1. The product of 5 and 3 is ____. For this reason, 5 and 3 are _____ of 15. (Factors are usually restricted to _____ numbers.)

   15
   factors
   natural

8   Chapter 1 The Real Number System

Find all natural number factors of each number.

2. 6 _____   1, 2, 3, 6

3. 18 _____   1, 2, 3, 6, 9, 18

4. 100 _____   1, 2, 4, 5, 10,

                                                 20, 25, 50, 100

5. The number 5^3 means that 5 is used as a factor
    ___ times. Find the value of 5^3.   3
    5^3 = ___ · ___ · ___ = ___   5; 5; 5; 125

6. In 5^3, the 3 is called the _____ and 5 is   exponent
    the _____.   base

Find the value of each of the following.

7. 4^2 = ___ · ___ = ___   4; 4; 16
8. 6^4 = _____   1296
9. 8^2 = _____   64
10. 9^3 = _____   729
11. $\left(\frac{3}{5}\right)^2$ = _____ $\frac{9}{25}$
12. $\left(\frac{9}{10}\right)^3$ = _____ $\frac{729}{1000}$

13. When more than one operation is used in a problem,
    use the following _____ of operations.   order
    If _____ _____ are present, simplify   grouping symbols
    within them, innermost first, and above and
    below _____ bars, separately, in the   fraction
    following order.
    (1) Apply all _____.   exponents
    (2) Do any multiplications or _____ in the   divisions
        order in which they occur, working from
        _____ to right.   left

1.2 Exponents, Order of Operations, and Inequality

(3) Do any _____ or subtractions in the order in which they occur, working from _____ to right.	additions left

14. What does $4(3 + 2)$ equal? First add inside parentheses: $4(3 + 2) = 4(\ \ )$. Then multiply: $4(5) = \underline{\ \ \ }$.

    5
    20

15. If an exercise involves more than one operation, first do the operations inside parentheses, as you did in Frame 14.

    $2(3 + 1) = 2(\ \ ) = \underline{\ \ \ }$ 4; 8
    $(4 - 2) + 3 = (\ \ ) + 3 = \underline{\ \ \ }$ 2; 5
    $5(7 + 1)(2) = 5(\ \ )(2) = \underline{\ \ \ } \cdot 2 = \underline{\ \ \ }$ 8; 40; 80

16. Do any multiplications or divisions before you do additions or subtractions. Go from let to right.

    $2 \cdot 1 + 4 + 10 \div 2 = \underline{\ \ \ } + 4 + \underline{\ \ \ }$ 2; 5
    $= \underline{\ \ \ }$ 11
    $3 + (5 - 1) + 3 \cdot 2 = 3 + \underline{\ \ \ } + \underline{\ \ \ }$ 4; 6
    $= \underline{\ \ \ }$ 13

17. Do additions and subtractions last, again from left to right.

    $4 + 2 + 5 \cdot 3 + 2(6 + 1)$
    $= 4 + 2 + \underline{\ \ \ } + 2(\ \ )$ 15; 7
    $= 4 + 2 + 15 + \underline{\ \ \ } = \underline{\ \ \ }$ 14; 35

18. When you have a fraction like

    $$\frac{2(6 + 1) + 14}{5 + 3 \cdot 3}$$

    first simplify the top (numerator) and bottom (denominator) of the fraction.

Chapter 1 The Real Number System

$$\frac{2(6 + 1) + 14}{5 + 3 \cdot 3}$$

$$= \frac{2(\quad) + 14}{5 + 3 \cdot 3} = \frac{\quad + 14}{5 + 3 \cdot 3} = \frac{\quad}{5 + 3 \cdot 3} \qquad 7; \ 14; \ 28$$

$$= \frac{28}{5 + \quad} = \frac{28}{\quad} \qquad\qquad 9; \ 14$$

Last of all, divide the numerator by the denominator.

$$\frac{28}{14} = \underline{\quad} \qquad\qquad 2$$

Simplify the following fraction

$$\frac{5(2 + 3) + 3}{2(1 + 1) + 3} = \frac{5(\quad) + 3}{2(\quad) + 3} = \frac{\quad + 3}{\quad + 3} = \underline{\quad} \qquad \begin{array}{l} 5; \ 25; \ 28 \\ 2; \ 4; \ 7 \end{array}$$

$$= \underline{\quad} \qquad\qquad 4$$

Simplify in Frames 19–26 by doing operations in the order you practiced in Frames 15–18. Work first inside parentheses, use any exponents, then do multiplications, then do additions or subtractions. Write fractions in lowest terms.

19. $4(2 + 7) = 4(\quad) = \underline{\quad}$ 9; 36

20. $3(4 + 5) = 3(\quad) = \underline{\quad}$ 9; 27

21. $2 \cdot 5 + 3 \cdot 4 = \underline{\quad} + \underline{\quad} = \underline{\quad}$ 10; 12; 22

22. $8 \cdot 9 + 4 = \underline{\quad} + 4 = \underline{\quad}$ 72; 76

23. $6(4 + 2^3) + 3 \cdot 5 = 6(4 + \underline{\quad}) + \underline{\quad}$ 8; 15
    $= 6(\underline{\quad}) + 15 = \underline{\quad} + \underline{\quad}$ 12; 72; 15
    $= \underline{\quad}$ 87

24. $2(9 + 2) + 6 \cdot 3 = \underline{\quad} + \underline{\quad} = \underline{\quad}$ 22; 18; 40

25. $\dfrac{6(5 + 9) - 3}{3(12 - 3)} = \dfrac{6(\quad) - 3}{3(\quad)} = \dfrac{\quad - 3}{\quad} = \underline{\quad}$ 14; 84; 81
                                                                           9; 27; 27
    $= \underline{\quad}$ 3

1.2 Exponents, Order of Operations, and Inequality

26. $\dfrac{4(9+1) + 4 \cdot 2}{3(2+3) + 1} = \dfrac{4(\underline{}) + \underline{}}{3(\underline{}) + 1} = \underline{}$

    $= \underline{}$

    10; 8; 48
    5; 16
    3

27. To avoid confusion by the use of double parentheses, use square _____ instead.

    brackets

Use the order of operations given above to find the values of the following. The parentheses are the innermost grouping symbols, so do the work inside them first.

28. $4[(2 \cdot 3) + (6 \cdot 2)] + 2 = 4[(\underline{}) + (\underline{})] + 2$
    $= 4(\underline{}) + 2 = \underline{}$

    6; 12
    18; 74

29. $(2+1)[(3 \cdot 4) + (2+2)] = (\underline{})[(\underline{}) + (\underline{})]$
    $= (\underline{})(\underline{}) = \underline{}$

    3; 12; 4
    3; 16; 48

30. $[(3 \cdot 2) + (2 \cdot 4)][(4 \cdot 3) - (2 + 1)]$

    $= \underline{}$

    $= \underline{}$

    $= \underline{}$

    $(6 + 8)(12 - 3)$
    $14(9)$
    126

31. $5[4(7-2) - 8] + (5-3)[19 - 2(3+6)]$

    $= \underline{}$

    62

32. Write $\dfrac{8(3+2) - 5}{7 + 10}$ using brackets and parentheses rather than a fraction bar.

    _____

    Simplify this expression as _____.

    $[8(3+2) - 5] \div$
    $(7 + 10)$
    $\dfrac{35}{17}$

33. The symbol ___, means "is not equal to." For example,

    $5 + 2 \underline{} 9$

    indicates that _____ is not equal to ___.

    $\neq$

    $\neq$
    $5 + 2;\ 9$

Chapter 1 The Real Number System

34. The symbol for "is less than" is <. Write "5 is less than 9" _____. Write "12 is less than 18" _____; "30 is less than 30 1/2" _____. The point of the symbol < points to the (*larger/smaller*) number.

    5 < 9
    12 < 18
    30 < 30 1/2
    smaller

35. The symbol for "is greater than" is ___. Write "10 is greater than 4" _____; "9 is greater than 5" _____; "18 is greater than 12" _____. The point of the symbol > points to the (*larger/smaller*) number.

    >
    10 > 4
    9 > 5
    18 > 12
    smaller

36. Insert < or > in each blank to make a true statement.

    9 ___ 5
    3 ___ 12
    1/2 ___ 1/3
    $2\frac{4}{5}$ ___ $2\frac{7}{10}$

    >
    <
    >
    >

37. "Is less than or equal to" is written with the symbol ___.
    Write "18 is less than or equal to 18" as _____. This statement is true because 18 is _____ to 18.

    ≤

    18 ≤ 18
    equal

38. "Is greater than or equal to" is written with the symbol ≥.
    Write "12 is greater than or equal to 7" as _____. This statement is true because 12 is _____ than 7. Write "11 is greater than or equal to 11" as _____. This is because 11 is _____ to 11.

    12 ≥ 7
    greater
    11 ≥ 11
    equal

1.2 Exponents, Order of Operations, and Inequality

39. Decide whether each statement is (*true/false*).

$9 \le 10$ _____	$15 < 15$ _____	true; false
$9\frac{1}{2} > 9$ _____	$16 \ge 8$ _____	true; true
$16 \ge 17$ _____	$0 < 5$ _____	false; true

40. Is $5 \le 8 + 2$ *true* or *false*? First add $8 + 2$: _____. Then $5 \le 8 + 2$ becomes $5 \le$ _____. This is (*true/false*). In the statements below, first add or subtract. Then decide whether each is *true* or *false*.

10; 10
true

$12 > 4 + 10$
$12 >$ _____ (*true/false*)   14; false

$6 + 2 + 5 \le 9 + 5$
_____ $\le$ _____ (*true/false*)   13; 14; true

$6 = 4 + 3$
$6 =$ _____ (*true/false*)   7; false

$8 - 2 > 4 + 1$
_____ $>$ _____ (*true/false*)   6; 5; true

41. In order to solve problems using mathematics, you must be able to translate the word statements of the problems into _____.

symbols

Write each statement in symbols.

42. Six is more than the product of four and one.

$6 > 4 \cdot 1$

43. Eleven is less than the quotient of thirty-four and three. _____

$11 < 34/3$

44. Twenty does not equal thirty. _____

$20 \ne 30$

45. Seven is greater than the product of two and three _____

$7 > 2 \cdot 3$

14   Chapter 1   The Real Number System

46. Any statement with < may be converted to one with ____, and any statement with > may be converted to one with ____.

>
<

47. For example, 5 > 3 says that 5 is _____ than 3. It is equally correct to say that 3 is ____ than 5, or 3 ____ 5.

greater

less; <

Write each statement with the inequality reversed.

48. 10 < 12 _____   12 > 10

49. 9 ≥ 2 _____   2 ≤ 9

50. 1/2 < 5/6 _____   5/6 > 1/2

51. .972 ≤ .973 _____   .973 ≥ .972

1.3 Variables, Expressions, and Equations

[1] Define variable, and find the value of an algebraic expression, given the values of the variables. (See Frames 1–11 below.)

[2] Convert phrases from words to algebraic expressions. (Frames 12–19)

[3] Identify solutions of equations. (Frames 20–23)

[4] Identify solutions of equations from a set of numbers. (Frames 24–26)

[5] Distinguish between an *expression* and an *equation*. (Frames 27–31)

1. A letter such as x represents a number whose value is not known. x is a variable. Letters such as x, y, z, a, b, c, are all examples of _____. In p + 11, the variable is ____.

variables
p

1.3 Variables, Expressions, and Equations   15

2. q + 10 combines a _____, a symbol for an operation, and a number. q + 10 is an algebraic expression. Then x + 11, 2p + 3, and x − 2y are all examples of algebraic _____. Write an expression showing that x is added to 4: _____;
Write an expression for "x minus 1": _____.

   variable

   expressions
   x + 4
   x − 1

3. Write the product of 9 and x without using a dot: ____. The product of 9 and x is 9x, the product 6 and p is _____, the product of 2 and y is _____. 11k, 3t, and 20m are all products of a number and a _____.

   9x
   6p; 2y

   variable

4. Suppose that x has the value 15. To find the value of x + 11, replace x with ____:
   x + 11 = ____ + 11. Then add the numbers to get ____. If x = 30, then x + 11 has the value ____ + 11 = ____. Find the value of x + 11 when x has the values 10, then 12, then 6 1/2.

   15
   15
   26
   30; 41

   If x = 10,   If x = 12,

   x + 11 = ____ + 11   x + 11 = ____ + 11
          = ____.   = ____.

   10; 12
   21; 23

   If   x = $6\frac{1}{2}$,

   x + 11 = ____ + 11 = ____.

   $6\frac{1}{2}$; $17\frac{1}{2}$

5. To find the value of 6p when p = 2, replace p with ____. Then multiply. 6p = 6( ) = ____.
   If p = 4, then 6p = 6( ) = ____. Find the value of 6p when p takes on the values 7 and 10.

   2; 2; 12
   4; 24

          If p = 7,   If p = 10,
            6p = ____.   6p = ____.

   42; 60

6. Find the value of 8x + 6y if x = 4 and y = 3.
       8x + 6y = 8( ) + 6( ) = ____ + ____
              = ____.

   4; 3; 32; 18
   50

Chapter 1 The Real Number System

7. Find the value of $2x - 4y + 8$ if $x = 7$ and $y = 2$.

    $2x - 4y + 8 = 2() - 4() + 8$ 7; 2

    $ = \underline{} - \underline{} + 8 = \underline{}$ 14; 8; 14

8. Find the value of $m^2 + 2n^2$ if $m = 3$ and $n = 4$.

    $m^2 + 2n^2 = ()^2 + 2()^2 = \underline{} + 2()$ 3; 4; 9; 16

    $ = \underline{} + \underline{} = \underline{}$ 9; 32; 41

Find the value of the following when $x = 7$.

9. $\dfrac{(3x - 7)(x + 2)}{(2x - 6)} = \underline{}$ $\dfrac{126}{8} = \dfrac{63}{4}$

10. $\dfrac{2(2x + 7 - 3x)}{x + 1} = \underline{}$ 0

11. $\dfrac{3x - 2x + 6}{x + 6} = \underline{}$ 1

12. Variables are used to convert word phrases to _____ expressions.   algebraic

Change word phrases to symbols in the following frames. Let x be the variable. Write symbols for any additions, subtractions, multiplications, or divisions.

13. 11 subtracted from a number

    11 subtracted from _____ $x - \underline{}$ x; 11

14. a number subtracted from 15

    _____ subtracted from 15   15 _____   x; − x

15. 7 times a number

    7 times _____   7( ) or _____   x; x; 7x

16. 7 times a number added to 70

    7 times _____ added to 70   70 + _____   x; 7x

1.3 Variables, Expressions, and Equations

17. the product of 8 and a number

    the product of 8 and ____   8( ) or ____ | x; x; 8x

18. a number divided by 4

    ____ divided by 4   ___ ÷ 4 or $\frac{}{4}$ | x; x; $\frac{x}{4}$

19. 6 times a number, divided by 9

    ____ divided by 9 $\frac{}{9}$ | 6x; $\frac{6x}{9}$

20. The statement 6p = 600 says that 6p and 600 are _____. The statement 6p = 600 is called an equation. Then 4x = 12, k + 2 = 30, and 3y − 2 = 19 are all examples of _____. Suppose the two expressions 6x and x + 15 are the same number. Write an equation showing this: ____ = _____ | equal

    equations

    6x = x + 15

21. 2x + 3 = 15 is an (*equation/expression*). Suppose x = 6. Replace x with 6 in the equation. Then do the operations. | equation

    $\qquad$ 2x + 3 = 15
    $\qquad$ 2( ) + 3 = 15   *Let x = 6* | 6
    $\qquad$ _____ + 3 = 15   *Multiply* | 12
    $\qquad\qquad$ _____ = 15   *Add* | 15

    The result, 15 = 15, is a (*true/false*) statement. | true
    The value x = 6 led to a _____ statement, so 6 | true
    is a _____ of the equation 2x + 3 = 15. | solution

22. Is 9 a solution of the equation 2x + 3 = 15? To find out, replace x with ____. | 9

    $\qquad$ 2x + 3 = 15
    $\qquad$ 2( ) + 3 = 15 | 9
    $\qquad$ _____ + 3 = 15 | 18
    $\qquad\qquad$ _____ = 15. | 21

18 Chapter 1 The Real Number System

The result is (*true/false*), so that 9 (*is/is not*) a solution of the equation $2x + 3 = 15$.	false is not
23. Is 3 a solution of the equation $5x - 4 = 11$? (*yes/no*)	yes
24. A list of numbers is sometimes written as a ____. Write the list of numbers 8, 9, 10, and 11 as a set. _____	set $\{8, 9, 10, 11\}$

Find all solutions for the following equations from the set $\{0, 2, 4, 6, 8, 10, 12\}$.

25. $5x + 3 = 43$ ____	8
26. $9p - 11 = 25$ ____	4
27. To distinguish between an expression and an equation, remember that an _____ is a phrase and an _____ is a sentence.	expression equation

Decide whether each of the following is an equation or an expression.

28. $4x + 2y = 6$	equation
29. $5x - 14y + 2$	expression
30. $\dfrac{3x - 6y}{14}$	expression
31. $x^2 + 2x + 1 = 0$	equation

1.4 Real Numbers and the Number Line

[1] Set up number lines. (See Frames 1-2 below.)

[2] Identify natural numbers, whole numbers, integers, rational numbers, irrational numbers, and real numbers. (Frames 3-8)

[3] Tell which of two different real numbers is smaller. (Frames 9-11)

[4] Find additive inverses of numbers. (Frames 12-13)

[5] Find absolute values of real numbers. (Frames 14-19)

---

1. The point 0 is marked on the _____ line below. Now locate natural numbers where they belong on this line.

   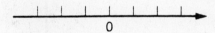

   number

   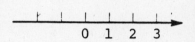

2. All numbers to the left of 0 on the number line are _____ numbers. All numbers to the right of 0 are _____ numbers.

   negative
   positive

3. The numbers that correspond to all points on a number line are called the _____ numbers. The set of real numbers contains these subsets: the natural numbers, the whole numbers, the integers, the rational numbers, and the irrational numbers.

   real

4. The set of whole numbers is included in the set of real numbers. The set of whole numbers is a _____ of the set of real numbers.

   subset

   Whole numbers: {___, 1, ___, ___, ___, ...}

   0; 2; 3; 4

   The three dots mean that the whole numbers (*continue/stop*). The next three whole numbers are ___, ___, and ___.

   continue
   5; 6; 7

Chapter 1 The Real Number System

Is 56 a whole number? (*yes/no*)	yes
Is -2 a whole number? (*yes/no*)	no
Is 3/4 a whole number? (*yes/no*)	no
Is 0 is a whole number? (*yes/no*)	yes

5. The natural numbers are a subset of the whole numbers. The whole numbers, in turn, are a _____ of the set of integers. Then, the integers are a _____ of the set of real numbers.

   subset
   subset

   Integers:

   $\{\ldots, \underline{}, \underline{}, \underline{}, 0, \underline{}, \underline{}, \underline{}, \ldots\}$

   -3; -2; -1;
   1; 2; 3

Is 56 an integer? (*yes/no*)	yes
Is -2 an integer? (*yes/no*)	yes
Is 3/4 an integer? (*yes/no*)	no

   Which of the following numbers are integers but not whole numbers?

   14, 0, -7, 362, 9, -9, -1058

   -7; -9; -1058

6. The integers together with fractions that are quotients of integers make up the set of rational numbers. (Division by zero is undefined.)

Is 56 a rational number? (*yes/no*)	yes
Is -2 a rational number? (*yes/no*)	yes
Is 3/4 a rational number? (*yes/no*)	yes
Is -5/9 a rational number? (*yes/no*)	yes
Is 8/0 a rational number? (*yes/no*)	no

7. Real numbers that are not rational numbers make up the set of irrational numbers, including square roots such as $\sqrt{2}, \sqrt{3}, \sqrt{5}$. The irrational numbers are a subset of the set of _____ numbers. The irrational numbers (*are/are not*) a subset of the set of rational numbers.

   real
   are not

1.4 Real Numbers and the Number Line   21

Is 56 an irrational number? (yes/no)	no
Is -2 an irrational number? (yes/no)	no
Is 3/4 an irrational number? (yes/no)	no

8. On the number line below, show where the negative numbers -2, -3, -4, and -5 fall.

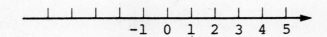   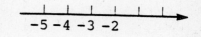

The numbers to the _____ of zero are called negative numbers. The three dots show that the negative numbers (*continue/stop*). Numbers to the right of zero on the number line are called _____ numbers. The three dots show that the positive numbers (*continue/stop*). Zero is neither positive nor negative. (*true/false*)

left

continue

positive
continue

true

9. Which of these two numbers on the number line is smaller?

The smaller number is ____. In symbols, -4 ___ -2.   -4; <
When two numbers are compared using the number line, the number to the (*left/right*) is smaller.   left

10. Write < or > so that the resulting statement is true. Imagine the numbers on a number line.

-6 ___ -4	0 ___ -6	-45 ___ 45
-3 ___ 1	-5 ___ -4	2/3 ___ 1/2
5 ___ -2	$1\frac{1}{2}$ ___ 0	3.1416 ___ 3.14159

<; >; <
<; <; > (4/6 > 3/6)
>; >; >

Chapter 1 The Real Number System

11. The number 0 is (*larger/smaller*) than any positive number.

    smaller

    The number 0 is (*larger/smaller*) than any negative number.

    larger

    Arrange the following numbers from smallest to largest.

    19, −3, 0, −15, 27, −45, 1/2

    smallest ___ ___ ___ ___ ___ ___ ___ largest

    −45; −15; −3; 0; 1/2; 19; 27

12. On the number line in Frame 8, the numbers 1 and −1 are the (*same/different*) distance from zero. The numbers 1 and −1 are on the (*same/opposite*) side of zero. Then 1 and −1 are additive inverses of each other. The numbers 6 and −6 are the _____ distance from zero, and on _____ sides of zero. 6 and −6 are _____ inverses of each other. −18 and 18 are another pair of additive _____. The additive inverse of 0 is ___.

    same
    opposite

    same; opposite
    additive

    inverses
    0

13. Write the additive inverse for each number.

Number	Additive inverse	Number	Additive inverse
12	_____	−7	_____
−4	_____	−(−2)	_____
−17	_____	$6\frac{1}{2}$	_____
0	_____	−(−12)	_____
4	_____	−8	_____

    −12; 7
    4; −2
    17; $-6\frac{1}{2}$
    0; −12
    −4; 8

14. The distance on a number line from a number to 0 is called the _____ value of the number. | absolute
The absolute value of 4 is ___ and the absolute value of -4 is ____. In symbols, $|4|$ = ____ and $|-4|$ = _____. Then $-|-8|$ = $-(\ )$ = ____. | 4
 | 4; 4
 | 4; 8; -8
The absolute value of the variable x is written ___. | $|x|$

15. Find the absolute values.

$|-4|$ = ____ | 4
$|12|$ = ____ | 12
$|-12|$ = ____ | 12
$-|-3|$ = $-(\ )$ = ____ | 3; -3
$-|-15|$ = $-(\ )$ = ____ | 15; -15
$-|2|$ = ____ | -2
$-|4|$ = ____ | -4

16. Decide whether the statement is *true* or *false*.

$-4 \geq |-3|$ _____ | false
$-9 \leq |-4|$ _____ | true
$-|-5| > -|-8|$ _____ | true
$|-2| < 0$ _____ | false
$0 < -|-9|$ _____ | false
$-2 \geq |-4|$ _____ | false

For what values of x and y are the following statements true?

17. $|x| = -|y|$ _____ | $x = y = 0$

18. $|5| = |x|$ _____ | $x = 5; x = -5$

19. $|x| = -x$ _____ | $x \leq 0$

24 Chapter 1 The Real Number System

1.5 Addition of Real Numbers

1 Add two numbers with the same sign. (See Frames 1-6 below.)

2 Add positive and negative numbers. (Frames 7-12)

3 Use the order of operations with real numbers. (Frames 13-18)

4 Interpret words and phrases that indicate addition. (Frames 19-23)

5 Interpret gains and losses as positive and negative numbers. (Frames 24-28)

---

1. Add 3 and 2 by arrows, as below. The starting point is ____.

   0

   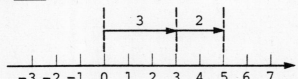

   The arrow for 3 goes ____ units to the _____. 3; right
   The arrow for 2 starts at ____ and goes ____   3; 2
   units to the _____. The ending point is ____.   right; 5
   Then 3 + 2 = ____.   5

2. The arrow over the number line below stands for
   the (*positive/negative*) number _____. The   negative; -2
   arrow starts at ____, ends at ____. It is ____   0; -2; 2
   units long, and goes to the (*right/left*).   left

   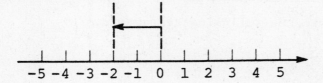

3. Add -1 and -3 by arrows over the number line
   below. -1 + (-3) = ____   -4

   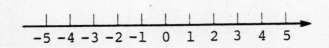   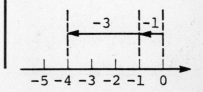

1.5 Addition of Real Numbers 25

In the sum −1 + (−3), do the parentheses mean
multiplication? (*yes/no*) Do they separate   no
the sign of a number and the operation symbol?
(*yes/no*)   yes

4. To add two numbers with the same ____, add the   sign
   _____ values of the numbers. The result   absolute
   has the _____ of the numbers being added.   sign

Add.

5. −17 + (−2) = ____   −19

6. −3 + (−19) = ____   −22

7. Add 6 and −3 by arrows.

           Arrow for 6:   for −3:
   Start   _____   _____   0; 6
   End   _____   _____   6; 3
   Direction _____   _____   right; left

   The sum 6 + (−3) = ____.   3

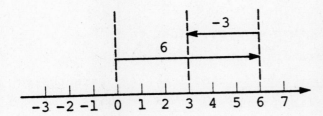

8. Use the number line to find the sums.

       3 + (−5) = ____   −2

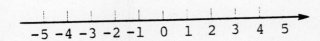

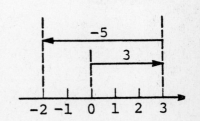

-5 + 8 = _____

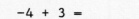

-4 + 3 = _____

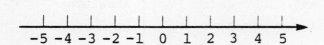

3

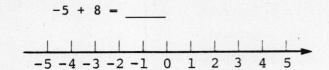

-1

9. To add numbers with unlike _____, find the _____ of the absolute values of the numbers. The answer has the sign of the number with the _____ absolute value.

signs
difference

larger

Add.

10. -7 + 5 = _____   -2

11. 28 + (-33) = _____   -5

12. -21 + 30 = _____   9

13. By the order of operations given earlier, addition problems with several numbers are worked by adding inside the brackets or _____ until a single number is obtained. Then add from _____ to _____.

parentheses

left
right

Use the order of operations to work the following problems.

14. (-4 + 6) + [-5 + (-6)]   Answer: _____   (2) + (-11) = -9

15. [-3 + (-2)] + (-4 + 8)   _____   -1

1.5 Addition of Real Numbers

16. [−4 + (−2)] + [−3 + (−2)]   _____   −11

17. (−4 + 8) + [−3 + (−9)] + 15   _____   7

18. −4 + [(−3 + 7) + (−9 + 12)]   _____   3

19. To work word problems with addition you must _____ words or phrases into symbols.   translate

Fill in the following table.

Word or Phrase	Example	Numerical Expression and Simplification	
20. Sum	the sum of −5 and 3	_____	−5 + 3 = −2
21. Added to	−7 added to −3	_____	−7 + (−3) = −10
22. More than	17 more than −4	_____	−4 + 17 = 12
23. Increased by	−9 increased by 3	_____	−9 + 3 = −6

24. Gains or increases in stated problems may be interpreted as _____ numbers. Losses or decreases are interpreted as _____ numbers.   positive   negative

For each problem, write a sum of real numbers and add. Then, write the answer to the problem.

25. Meredith earned $38 as a library aide. She spent $16 of it at the book store. How much did she have left?

    _____ = _____   38 + (−16); 22

    Answer: _____ $22

Chapter 1 The Real Number System

26. Carlos deposited $75 in his checking account and wrote a check for $110. Find the amount more or less in Carlos's checking account as a result of the two transactions.

　　　　　_____ = _____

　　　　　　　Answer: _____

75 + (−110); −35

$35 less

27. On the first down, Mike Pagel was sacked for a loss of 5 yards. On the second down, Pagel passed to Kevin Mack for a gain of 11 yards. What was the net gain or loss for the two downs?

　　　　　_____ = _____

　　　　　　　Answer: _____

−5 + 11; 6

6 yards gain

28. On January 10, the temperature at 7 P.M. in Ridgway, Pennsylvania, was −2°F. By 3 A.M. the next morning the temperature had decreased by 11°. What was the temperature at 3 A.M.?

　　　　　_____ = _____

　　　　　　　Answer: _____

−2 + (−11); −13

−13°F.

1.6 Subtraction of Real Numbers

[1] Find a difference on a number line. (See Frames 1 below.)

[2] Use the definition of subtraction. (Frames 2–6)

[3] Work subtraction problems that involve grouping. (Frames 7–18)

[4] Interpret words and phrases that involve subtraction. (Frames 19–22)

[5] Solve problems that involve subtraction. (Frames 23–27)

1. To find 8 − 5 on a number line, start at _____ and draw an arrow _____ units to the (*right/left*). From the right end of this arrow, draw an arrow _____ units to the (*right/left*).

0

8; right

5; left

1.6 Subtraction of Real Numbers

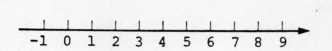

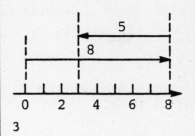

Answer: 8 − 5 = ____   3

2. As a definition, to subtract two numbers, change the ____ on the second number and ____ .   sign; add

3. In symbols: x − y = _____ .   x + (−y)

4. Subtract −2 from 3. Write 3 − ( ). The parentheses separates the sign of the number and the operation _____ .   −2

   symbol

   Step 1: Change the sign of −2.
   3 − (−2) = 3 + ( )   2

   Step 2: Add.
   = ____   5

   Again: 3 − (−2) = 3 _____ = ____ . But   + (2); 5
   3 + (−2) = _____ .   1

5. Subtract −1 from −4: −4 − (−1) = −4 + (1) = ____ .   −3
   Subtract 8 from 6: 6 − 8 = 6 ____ (−8) = ____ .   +; −2
   Here we change 8 to −8 and add.
   Subtract 3 from −6: −6 − 3 = −6 ____ ( ) = ____ .   + (−3); −9

6. Change signs and add.
   4 − (−9) = 4 + ( ) = ____   9; 13
   −9 − (−2) = −9 ____ ( ) = ____   + (2); −7
   −5 − (−8) = −5 + ( ) = ____   8; 3
   3 − 5 = 3 ____ ( ) = ____   + (−5); −2
   11 − 15 = 11 + ( ) = ____   −15; −4
   −5 − 8 = −5 ____ ( ) = ____   + (−8); −13
   −6 − 4 = −6 + ( ) = ____   −4; −10
   4 − (−12) = 4 ____ ( ) = ____   + (12); 16

Chapter 1 The Real Number System

7. When a problem involves grouping symbols, work _____ the grouping symbols first. | inside

Work each problem in Frames 8–18.

8. $1 + [-6 - (-2)] = 1 + (-6 + \quad)$
  $= 1 + (\quad) = \underline{\quad}$ | 2
 -4; -3

9. $[-2 - (-5)] + (-8) = [-2 + (\quad)] + (\quad)$
  $= \underline{\quad} + (-8) = \underline{\quad}$ | 5; -8
 3; -5

10. $[-4 - (-3)] - (-2) = \underline{\quad} - (-2)$
  $= \underline{\quad} + \underline{\quad} = \underline{\quad}$ | -1
 -1; 2; 1

11. $(4 - 9) - [3 - (-2)] = \underline{\quad}$ | -10

12. $(-3 + 4) - [-2 - (-5)] = \underline{\quad}$ | -2

13. $(-5 + 9) - [-3 - (-1)] = \underline{\quad}$ | 6

14. $[-5 - (-6)] - [-4 - (-2)] = \underline{\quad}$ | 3

15. $[-6 - (-4)] - [-3 - (-2)] = \underline{\quad}$ | -1

16. $[-4 - 3] - [-2 - (-1)] = \underline{\quad}$ | -6

17. $[-5 - (-2)] - [-3 - (-6)] = \underline{\quad}$ | -6

18. $[-9 - 4] - [3 - (2 - 4)] = \underline{\quad\quad\quad\quad}$ | $(-13) - [3 - (-2)]$
 $= (-13) - (3 + 2)$
 $= (-13) + (-5)$
 $= -18$

1.6 Subtraction of Real Numbers

Fill in the following table of key words and phrases that indicate subtraction.

	Word or Phrase	Example	Numerical Expression and Simplication	
19.	Difference	The difference between −5 and −6	_____ _____	−5 − (−6) = −5 + 6 = 1
20.	Subtracted from	14 subtracted from 8	_____ _____	8 − 14 = 8 + (−14) = −6
21.	Less than	10 less than 2	_____ _____	2 − 10 = 2 + (−10) = −8
22.	Decreased by	4 decreased by −2	_____ _____	4 − (−2) = 4 + 2 = 6

23. Gains, profits in business, temperatures above 0°, and altitudes above sea level may be interpreted as _____ numbers. Losses, temperatures below 0°, and altitudes below sea level may be interpreted as _____ numbers.

positive

negative

For each problem, write a difference of real numbers and subtract. Then, write the answer to the problem.

24. Marlo read a gauge in the lab that said −4. An hour later the reading was 6 less. Find the second reading.

_____ = _____

Answer: _____

−4 − 6; −10

−10

Chapter 1 The Real Number System

25. Laurel Summit, Pennsylvania, has an altitude of 2743 feet above sea level. Death Valley, California, has an altitude of 282 feet below sea level. Find the difference between the two elevations.

    _____ - _____   2743 - (-282); 3025

    Answer: _____   3025 feet

26. Diamond Brothers own two companies, Valutek and Producta. Last month Valutek showed a profit of $9050, while Producta lost $1500. Find the difference in the amounts earned by the two companies.

    _____ - _____   9050 - (-1500); 10,550

    Answer: _____ $10,550

27. The temperature at Cleveland, Ohio, on December 9 was 5° colder than at Columbus, Ohio. If the temperature was -1°F at Columbus, find the temperature at Cleveland.

    _____ - _____   -1 - 5; -6

    Answer: _____   -6°

1.7 Multiplication of Real Numbers

[1] Find the product of a positive and a negative number. (See Frames 1-4 below.)

[2] Find the product of two negative numbers. (Frames 5-8)

[3] Identify factors of integers. (Frames 9-14)

[4] Use the order of operations in multiplication with signed numbers. (Frames 15-20)

[5] Evaluate expressions involving variables. (Frames 21-24)

[6] Interpret words and phrases that indicate multiplication. (Frames 25-29)

1.7 Multiplication of Real Numbers

1. The product of two numbers with unlike signs is _____.   negative

Find each product.

2. $(-5)(9) =$ ____   −45

3. $8(-7) =$ ____   −56

4. $(-12)11 =$ ____   −132

5. The product of two numbers with like signs is _____.   positive

Find each product.

6. $(-4)(-8) =$ ____   32

7. $\left(-\frac{1}{2}\right)(-6) =$ ____   3

8. The product $3 \cdot 0 = 0 \cdot 2\frac{1}{2} = 0$, $16 \cdot 0 =$ ____,   0
   $45 \cdot$ ____ $= 0$, $0(7 + 2) =$ ____, $9 \cdot 0 = 0$; and   0; 0
   $x \cdot 0 =$ ____, where x is any _____ number.   0; real
   Then $0(-2) =$ ____ and $-32(0) =$ ____, and so on.   0; 0
   The product of 0 and any real number is ____.   0

9. You have learned the definition of factors for whole numbers. The definition may be extended to integers. If the product of two integers is a third integer, each of the two integers is a _____ of the third.   factor

Find all integer factors of each number.

10. 9   _____   −9, −3, −1, 1, 3, 9

34   Chapter 1   The Real Number System

11. -16   _____   -16, -8, -4, -2,
                                                      -1, 1, 2, 4, 8,
                                                      16

12. -3   _____   -3, -1, 1, 3

13. 24   _____   -24, -12, -8, -6,
                                                      -4, -3, -2, -1, 1,
                                                      2, 3, 4, 6, 8, 12,
                                                      24

14. 11   _____   -11, -1, 1, 11

Use the order of operations to find each of the following.

15. (-4 + 3)(-5 - 6) = (   )(   ) = _____   (-1)(-11); 11

16. (-8 + 2)(3) + 4 = (   )(3) + 4   -6
                     = _____ + 4   -18
                     = _____   -14

17. (-4 + 2)(-3 + 5) + 6 = (   )(   ) + 6   (-2)(2)
                          = _____ + 6 = ____   -4; 2

18. To work $|-3(-4)| + (-5)(2)$, work inside the
     absolute value bars the same way you work
     inside parentheses.

     $|-3(-4)| + (-5)(2)$ = | | + ___   12; (-10)
                          = ____ + ____   12; (-10)
                          = ____   2

1.7 Multiplication of Real Numbers   35

19. $[-2(-3) + 5](7 - 3) + |4(-6)|$
    $= (+ 5)() + ||$ 6; 4; −24
    $= ()() + \underline{}$ 11; 4; 24
    $= \underline{} + \underline{} = \underline{}$ 44; 24; 68

20. $(-4 - 7)(-9 + 6) - (-8)(2) = \underline{}$ 49

In each of the following, evaluate the expression by replacing x with −4 and y with 6.

21. $3x + 4y = 3() + 4()$ −4; 6
    $= \underline{} + \underline{} = \underline{}$ −12; 24; 12

22. $9x - 6y + 4 = \underline{}$ −68

23. $(3 + x)(5 + y) = \underline{}$ −11

24. $2x^2 - y^2 = 2()^2 - ()^2$ −4; 6
    $= 2() - \underline{}$ 16; 36
    $= \underline{} - \underline{} = \underline{}$ 32; 36; −4

Fill in the following table.

	Word or Phrase	Example	Numerical Expression and Simplification	
25.	Product	The product of −4 and −1	_____	$(-4)(-1) = 4$
26.	Times	17 times −8	_____	$17(-8) = -136$
27.	Twice	Twice 3	_____	$2(3) = 6$
28.	Of (used with fractions)	$\frac{1}{3}$ of 12	_____	$\frac{1}{3}(12) = 4$
29.	Percent of	15% of −45	_____	$.15(-45) = -6.75$

Chapter 1 The Real Number System

1.8 Division of Real Numbers

[1] Find the reciprocal, or multiplicative inverse, of a number. (Frames 1-6 below.)

[2] Divide using signed numbers. (Frames 7-10)

[3] Simplify numerical expressions involving quotients. (Frames 11-18)

[4] Interpret words and phrases that indicate division. (Frames 19-20)

[5] Translate simple sentences into equations. (Frames 21-24)

---

1. Pairs of numbers whose product is ____ are called multiplicative _____; or _____ of of each other.

   1 inverses; reciprocals

Find the multiplicative inverse of each number.

2. 7 _____

   $\frac{1}{7}$

3. -9 _____

   $-\frac{1}{9}$

4. $\frac{5}{8}$ _____

   $\frac{8}{5}$

5. $-\frac{9}{11}$ _____

   $-\frac{11}{9}$

6. 0 _____

   none

7. The quotient of two negative numbers is (*negative/positive*). Then $\frac{-4}{-2}$ = _____.

   positive; 2

8. The quotient of a negative and a positive number and the quotient of a positive and a negative number are (*negative/positive*). Then $\frac{-4}{2}$ = _____ and ____.

   negative; -2
   -2

1.8 Division of Real Numbers

9. The symbol $\frac{2}{0}$ (does/does not) represent a _____ number. Division by zero is _____.

   Is $\frac{8}{0}$ a real number? (yes/no)

   does not; real
   undefined

   no

10. Find the quotients.

    $\frac{8}{-2} = $ _____ $\frac{-49}{7} = $ _____   −4; −7

    $\frac{6}{-3} = $ _____ $\frac{0}{5} = $ _____   −2; 0

    $\frac{12}{-4} = $ _____ $\frac{-18}{-2} = $ _____   −3; 9

    $\frac{-4}{2} = $ _____ $\frac{-27}{-3} = $ _____   −2; 9

    $\frac{-15}{3} = $ _____ $\frac{-100}{-5} = $ _____   −5; 20

    $\frac{0}{-6} = $ _____ $\frac{-180}{-15} = $ _____   0; 12

Simplify the numerators and denominators separately, and find the quotients.

11. $\frac{9(3+2)}{-3(7-2)} = \frac{9(\quad)}{-3(\quad)} = \frac{\quad}{\quad} = $ _____

    5; 45
    5; −15; quotient is −3

12. $\frac{-6-8}{3-(-4)} = \frac{\quad}{\quad} = $ _____

    −14
    7; quotient is −2

13. $\frac{-4(8)-3}{-7-(-2)} = \frac{\quad -3}{\quad} = \frac{\quad}{-5} = $ _____

    −32; −35
    −5; quotient is 7

14. $\frac{4(-2)+8(-3)}{-5(3)+(-1)}$ Answer: _____   2

15. $\frac{7(-3)-(-9)}{5(-3)-(-3)}$ _____   1

16. $\frac{4^2+3^2}{5(6-5)}$ _____   5

17. $\frac{4-(3^2+7)}{6(-2)-3(5)}$ _____ $\frac{4}{9}$

38   Chapter 1 The Real Number System

18. $\dfrac{-7(-3) - (5^2 - 4^2)}{3^2 - [5 + 7(4)]}$ _____ $-\dfrac{1}{2}$

Fill in the following table.

	Word or Phrase	Example	Numerical Expression and Simplication	
19.	Quotient	The quotient of 63 and −9	_____	$\dfrac{63}{-9} = -7$
20.	Divided by	−48 divided by 6	_____	$\dfrac{-48}{6} = -6$

Write each sentence in symbols and find the solution by guessing or trial and error. The solutions come from the set of integers between −12 and 12, inclusive.

21. Fifty-five divided by a number is −5.

   _____ $\dfrac{55}{x} = -5$

   Solution: _____ -11

22. The quotient of a number and −6 is −2.

   _____ $\dfrac{x}{-6} = -2$

   Solution: _____ 12

23. The square of 4 divided by a number is −8.

   _____ $\dfrac{4^2}{x} = -8$

   Solution: _____ -2

25. When the quotient of a number and 2 is added to −5, the result is −9.

   _____ $\dfrac{x}{2} + (-5) = -9$

   Solution: _____ -8

1.9 Properties of Addition and Multiplication

[1] Identify the use of the commutative properties. (See Frames 1-4 below.)

[2] Identify the use of the associative properties. (Frames 5-9)

[3] Identify the use of the identity properties. (Frames 10-11)

[4] Identify the use of the inverse properties. (Frames 12-27)

[5] Identify the use of the distributive property. (Frames 28-43)

---

1. Adding 4 + 9 gives you the same answer as adding 9 + ____. Then 4 + 9 = ____ + 4. If x and y any two real numbers, then ____ = y + x. This property of addition is the _____ property.

    4; 9
    x + y
    commutative

2. Complete the given statement so that it is an example of the commutative property for addition.

    ____ + 7 = 7 + 9   9
    4 + 11 = ____ + 4   11
    9 + (6 + 2) = (6 + 2) + ____   9
    (4 + 3) + 5 = (____ + 4) + 5   3
    (5 + 8) + (3 + 2) = (____) + (5 + 8)   3 + 2

3. By the commutative property of multiplication, the product $4 \cdot 2$ is the same as the product ____, or ____ = ____. In symbols, if x and y are any two real numbers, then by the commutative property, ____ = ____.

    $2 \cdot 4$; $4 \cdot 2$; $2 \cdot 4$

    xy; yx

4. Complete the given statement so that it is an example of the commutative property for multiplication.

    8(____) = $6 \cdot 8$ 6
    -4(____) = (-3)(-4)   -3
    (5 + 6)(2) = (____)(5 + 6)   2
    (-3 + 4)(____) = 5(-3 + 4)   5

Chapter 1 The Real Number System

5. To add 4, 6, and 8, you could group them either $(4 + 6) + 8$ or $4 + (6 + 8)$. Decide whether the statement $(4 + 6) + 8 = 4 + (6 + 8)$ is *true* or *false*. _____ You can associate (or group) the sum $5 + (7 + 9)$ in another way: $(\_ + \_) + 9$. This property of addition, that assures the same answer from either grouping, is the (*commutative*/ *associative*) property.

   true
   5; 7

   associative

6. Complete the given statement so that it is an example of the associative property of addition.

   $4 + (9 + 5) = (\_\_\_) + 5$ $4 + 9$
   $(6 + 2) + 11 = 6 + (\_\_\_)$ $2 + 11$
   $9 + [8 + (3 + 2)] = (\_\_\_) + (3 + 2)$ $9 + 8$
   $(x + y) + z = \_\_\_ + (y + z)$ x

7. By the associative property for multiplication, the product $4(3 \cdot 5)$ is the same as the product $(\_\_\_) \cdot 5$. In symbols, if x, y, and z are any real numbers, then the associative property says that $(xy)z = \_\_\_$.

   $4 \cdot 3$

   $x(yz)$

8. Complete the given statement so that it is an example of the associative property for multiplication.

   $6[2(\_\_\_)] = (6 \cdot 2)(-8)$ -8
   $-4(3 \cdot 5) = (-4 \cdot 3) \cdot \_\_\_$ 5
   $-8(9 \cdot 2) = (-8 \cdot 9) \cdot \_\_\_$ 2

9. Identify the statement as an example of either the commutative or the associative property.

   $8 + (-4) = -4 + 8$ _____   commutative
   $3(4x) = (3 \cdot 4)x$ _____   associative
   $y[4(-1)] = [4(-1)]y$ _____   commutative
   $3 + \frac{1}{2} = \frac{1}{2} + 3$ _____   commutative

1.9 Properties of Addition and Multiplication 41

$2 \cdot 5 + (3\frac{1}{2} + \frac{1}{3}) = (2 \cdot 5 + 3\frac{1}{2}) + \frac{1}{3}$ _____	associative
$x + (2x + 1) = (x + 2x) + 1$ _____	associative

10. Add zero: $9 + 0 =$ ____ and $0 + 12 =$ ____. The sum of any real number and _____ is that same real number. This property of addition is the _____ property. By the identity property, $9 +$ ____ $= 0 +$ ____ $= 9$ and $0 + 12 =$ ____ $+ 0 = 12$. The number zero is the _____ element for addition.

    9; 12
    zero

    identity
    0; 9; 12
    identity

11. By the identity property of multiplication, the product of any real number and ____ is the original real number. In symbols, if x is any real number, then the identity property says that $x \cdot 1 =$ ____ $\cdot$ ____ $=$ ____. The number ____ is called the identity _____ for multiplication.

    1

    1; x; x; 1
    element

12. The real numbers 8 and -8 are the (*same/different*) distance from 0 on the number line, and are located on the (*same/opposite*) side of 0. Then 8 and -8 are _____ inverses of each other.

    same

    opposite
    additive

13. What is the sum of 8 and -8? The sum $8 + (-8) =$ ____. The sum $-9 + 9 =$ ____. Then $x +$ ____ $= 0$ and ____ $+ x = 0$. The inverse property says that if x is any real number, then a real number $-x$ exists such that the sum of x and ____ is zero. In other words, the sum of additive ____ is zero.

    0; 0; -x
    -x

    -x
    inverses

14. The product of the numbers 4 and 1/4 is ____. For this reason, these numbers are called multiplicative _____, or _____, of each other.

    1

    inverses;
    reciprocals

Chapter 1 The Real Number System

In symbols, the multiplicative inverse property says that if x is any real number (except ____) then there is a real number 1/x such that x · ( ) = 1/x · ( ) = ____. Write the reciprocal of each number. (Assume $x \neq 0$.)

Number	Reciprocal	Number	Reciprocal
6	_____	17	_____
−3	_____	−3/4	_____
1/2	_____	−7	_____
1	_____	2/3	_____
3/4	_____	1/6	_____
11/18	_____	x	_____

0

1/x; x; 1

1/6; 1/17
−1/3; −4/3
2; −1/7
1; 3/2
4/3; 6
18/11; 1/x

15. Decide whether each fraction is an example of the (*identity/inverse*) property.

$12 + (-12) = 0$	_____
$1 \cdot x = x$	_____
$9 + 0 = 0$	_____
$-\frac{4}{3} + \frac{4}{3} = 0$	_____
$\left(-\frac{1}{8}\right)(-8) = 1$	_____
$1(2x + 3) = 2x + 3$	_____

inverse
identity
identity
inverse
inverse
identity

Decide if each statement is an example of the commutative, associative, identity, or inverse property.

16. $5 + 9 = 9 + 5$ _____ commutative

17. $(-6)1 = -6$ _____ identity

18. $3 + (8 + 2) = (3 + 8) + 2$ _____ associative

19. $(4 + 2) + 6 = 6 + (4 + 2)$ _____ commutative

20. $a(5 + 11) = (5 + 11)a$ _____ commutative

1.9 Properties of Addition and Multiplication

21. $4 + 0 = 4$ _____ identity

22. $3 + (-3) = 0$ _____ inverse

23. $-6 + [4 + (-5)] = -6 + (-5 + 4)$ _____ commutative

24. $-y\left(-\frac{1}{y}\right) = 1$ _____ inverse

25. $0 + (-14) = -14$ _____ identity

26. $3y(y + 1) = 3[y(y + 1)]$ _____ associative

27. $12 + [-12 + 12] = 12 + 0$ _____ inverse

28. You found the answer to $4(2 + 6)$ by first adding inside parentheses and then multiplying:
$4(2 + 6) = 4(\ \ \ ) = \underline{\ \ \ \ }$. Do this another way: 8; 32
$4(2 + 6) = 4 \cdot 2 + 4 \cdot 6 = \underline{\ \ \ } + \underline{\ \ \ } = \underline{\ \ \ }$. Both   8; 24; 32
ways give the (*same/different*) answer. Again: same
$4(2 + 6) = 4 \cdot \underline{\ \ \ } + 4 \cdot \underline{\ \ \ } = \underline{\ \ \ } + \underline{\ \ \ } = \underline{\ \ \ }$.   2; 6; 8; 24; 32
You distributed the 4 by using the _____   distributive
property $3(2 + 1) = 3 \cdot 2 + 3 \cdot \underline{\ \ \ } = \underline{\ \ \ } + \underline{\ \ \ } =$ 1; 6; 3
$\underline{\ \ \ }$. You distributed the 3.   9

29. Multiply each number inside parentheses by the outside number. Then add.

      Distribute   *Multiply*   *Add*

$4(9 + 2) = 4 \cdot 9 + 4 \cdot \underline{\ \ \ } = \underline{\ \ \ } + \underline{\ \ \ } = \underline{\ \ \ }$ 2; 36; 8; 44

$6(3 + 11) = 6 \cdot \underline{\ \ \ } + \underline{\ \ \ } \cdot 11 = \underline{\ \ \ } + \underline{\ \ \ } = \underline{\ \ \ }$ 3; 6; 18; 66; 84

$7(x + 1) = 7 \cdot \underline{\ \ \ } + 7 \cdot \underline{\ \ \ } = \underline{\ \ \ } + \underline{\ \ \ }$ x; 1; 7x; 7

$3(x - 4) = \underline{\ \ \ } \cdot \underline{\ \ \ } - \underline{\ \ \ } \cdot \underline{\ \ \ } = \underline{\ \ \ } - \underline{\ \ \ }$ 3; x; 3; 4; 3x; 12

Chapter 1 The Real Number System

30. It is true that 6(3 + 11) = (3 + 11)6 by the (*commutative/associative*) property. It is true that 6(3 + 11) = 6·3 + 6·11 and (3 + 11)6 = 3·6 + 11·6 by the (*commutative/distributive*) property. Complete each statement to make it an example of the distributive property.

    4(9 + 2) = 4·___ + 4·___
    6(3 + 11) = ___·___ + ___·___
    ___ (8 + 6) = 4·8 + 4·6
    ___ (2 + 4) = 5·2 + 5·4
    4(___ + 8) = 4·6 + 4·8

commutative

distributive

9; 2
6; 3; 6; 11
4
5
6

31. Simplify by getting rid of the parentheses. Use the distributive property:

    2(3 − 4a) + 5c = ___ − ___ + 5c.

6; 8a

32. Simplify 8(3 + 5b) − 4(2 − 3c).

    8(3 + 5b) − 4(2 − 3c)
    = 8·___ + 8·___ − ___·2 − ___ (___)
    = ___ + ___ − ___ + ___
    = _____

3; 5b; 4; 4; −3c
24; 40b; 8; 12c
16 + 40b + 12c

33. −2(3a + 2b − c)   _____   −6a − 4b + 2c

34. −(4m − 3k)   _____   −4m + 3k

Write the property illustrated by each given statement.

35. 8(9 + 4) = 8·9 + 8·4   _____   distributive

36. 4·8 = 8·4   _____   commutative

37. 9·(6·2) = 9·(2·6)   _____   commutative

38. $12\left(\frac{1}{12}\right) = 1$ _____   inverse

1.9 Properties of Addition and Multiplication

39. $8 \cdot 1 = 8$ _____ | identity

40. $(-4)\left(-\frac{1}{4}\right)$ _____ | inverse

41. $6 \cdot 2 + 6 \cdot 9 = 6(2 + 9)$ _____ | distributive

42. $1 \cdot 12 = 12$ _____ | identity

43. $6(2 \cdot 5) = (6 \cdot 2) \cdot 5$ _____ | associative

46   Chapter 1   The Real Number System

Chapter 1 Test

The answers for these questions are at the back of this Study Guide.

1. Write $\frac{45}{120}$ in lowest terms.   1. _____

2. Add: $\frac{1}{4} + \frac{5}{9} + \frac{11}{12}$.   2. _____

3. Divide: $\frac{7}{3} \div \frac{5}{18}$.   3. _____

First simplify each side of the inequality symbol and then answer *true* or *false* for the following.

4. $-|-2| - 7 \geq -5$ 4. _____

5. $2[3(-4) - (-6)(5)] \leq 0$ 5. _____

6. $\frac{-5(2) - 4(-3)}{-4(-8) - (-4)} > -1$ 6. _____

7. $\frac{9(5 - 2) + 3(1 - 4)}{4(5 - 7) + 2(4 - 8)} \geq 0$ 7. _____

8. $(-4)^2 - 2^2 \leq 6^2$ 8. _____

Find the numerical value of the given expression when $m = 3$ and $p = -4$.

9. $2m^2 + 3p$ 9. _____

10. $\frac{p^3 + 3m^2}{3 - p}$ 10. _____

Select the smaller number from each pair.

11. $-2, -|-3|$ 11. _____

12. .374, .3709   12. _____

Chapter 1 Test

Write each word phrase as an algebraic expression. Use x as the variable.

13. Ten less than 3 times a number   13. _____

14. The quotient of 4 and the sum of a number and 5   14. _____

Decide whether the given number is a solution of the equation.

15. $4x - x + 6 = 15$; 3   15. _____

16. $3(m + 6) - 7m = -2$; -5 16. _____

Perform the indicated operations. If the expression is undefined, say so.

17. $-1 + (7 - 2) - (-12)$ 17. _____

18. $-3\frac{1}{3} + 7\frac{3}{8}$ 18. _____

19. $-4 - [3 - (9 - 7)]$ 19. _____

20. $4^2 + (-4) - (7 - 3^2)$ 20. _____

21. $(-2) \cdot |-4| - |5| \cdot (4)$ 21. _____

22. $\dfrac{(3 - 6) - (2 - 8)}{-7 - (-5)}$ 22. _____

23. $\dfrac{2[-3 - (-2 + 5)]}{-4[-1 - (-3)] - 2(-5)}$ 23. _____

24. $\dfrac{5 - 2(-7 + 10)}{9(-4) + (-5 - 1)(-3 - 3)}$ 24. _____

Find the solution for each equation. Choose solutions from the domain $\{-7, -4, -1, 0, 1, 4, 7\}$.

25. $\dfrac{z}{2} = -2$ 25. _____

26. $3x + 4 = -17$ 26. _____

27. $-7y + 2 = 9$ 27. _____

Match the property in Column I with all examples of it from Column II.

      Column I   Column II

28. Commutative   A. $6z = z \cdot 6$ 28. _____

29. Associative   B. $4 + (3 + p) = (4 + 3) + p$ 29. _____

30. Identity   C. $4 + (-4) = -4 + 4$ 30. _____

31. Inverse   D. $-(5 - m) = -5 + m$ 31. _____

32. Distributive   E. $\frac{2}{3}\left(\frac{3}{2}\right) = 1$ 32. _____

                         F. $-8 + 0 = -8$

                         G. $1(-7) = -7$

                         H. $-3 + 3 = 0$

                         I. $3(7) + 3(4) = 3(7 + 4)$

33. Use the distributive property to
    simplify $-(6z - 9)$.   33. _____

CHAPTER 2   SOLVING EQUATIONS AND INEQUALITIES

2.1 Simplifying Expressions

[1] Simplify expressions.   (See Frames 1-4 below.)

[2] Identify terms and numerical cofficients.   (Frames 5-10)

[3] Identify like terms.   (Frames 11-15)

[4] Combine like terms.   (Frames 16-22)

[5] Simplify expressions from word phrases.   (Frames 23-30)

---

1. To solve an equation it is often necessary to first _____ the equation. We start by simplifying expressions. — simplify

Simplify each expression.

2. $2(7q - 8y) =$ _____ — $14q - 16y$

3. $8 + 3(4p + 3) = 8 +$ ____ $+$ ____ $=$ _____ — $12p$; 9; $12p + 17$

4. $1 - (8y - 2) =$ _____ — $3 - 8y$

5. The numerical _____ of $8r$ is _____. — coefficient; 8

6. A _____ is a single number, a variable, or a _____ or _____ of numbers and variables raised to powers. — term; product; quotient

Give the numerical coefficient for each term.

7. $7z$ _____ — 7

8. $-11y^4$ _____ — -11

9. m^5 _____ — 1

10. $-k^{12}$ _____ — -1

Chapter 2 Solving Equations and Inequalities

11. Terms with exactly the same variables and the same exponents are _____ terms.   —   like

Write like or unlike for each list of terms.

12. $12z$, $5z$ _____   —   like

13. $-4k^2$, k^2, $7k^2$ _____   —   like

14. $8m^5$, $7m^3$ _____   —   unlike

15. Only _____ terms may be combined. To do so, use the _____ property.   —   like; distributive

Combine like terms.

16. $7m + 9m = ($ _____ $)m =$ _____   — $7 + 9$; $16m$

17. $8k^4 + 7k^4 + 2k^4 =$ _____   — $17k^4$

18. $-3r^2 + 6r^2 + r^2 = -3r^2 + 6r^2 +$ _____ r^2
    = _____   — 1; $4r^2$

19. $8m^3 - 9m^2 =$ _____   —   Unlike terms cannot be combined.

20. $4(3z - 6) = 4($ _____ $) - 4($ _____ $)$
    = _____   — $3z$; 6; $12z - 24$

21. $8(5k - 7) + 4 =$ _____   — $40k - 52$

22. $(-6y^2 + 2) - (y^2 - 8) =$ _____   — $-7y^2 + 10$

2.1 Simplifying Expressions

Convert each word phrase to a mathematical expression. Use x as the variable. Combine terms whenever possible.

23. Five more than a number _____

    $5 + x$ or $x + 5$

24. Six times twice a number _____

    $6(2x)$ or $12x$

25. The quotient of a number and one more than the number _____

    $\frac{x}{1+x}$ or $\frac{x}{x+1}$

26. The difference between ten and a number, with the result subtracted from the sum of eight times the number and four _____

    $8x + 4 - (10 - x)$ or $9x - 6$

27. The product of eight and the sum of twice a number and five _____

    $8(2x + 5)$ or $16x + 40$

28. Half a number added to eleven _____

    $\frac{1}{2}x + 11$

29. The quotient of eight times a number and four, with the result subtracted from the sum of the number and six _____

    $(x + 6) - \frac{8x}{4}$ or $-x + 6$

30. The product of 8 and the quotient of a number and 5 _____

    $8\left(\frac{x}{5}\right)$

Chapter 2 Solving Equations and Inequalities

2.2 The Addition and Multiplication Properties of Equality

1. Identify linear equations. (See Frames 1-2 below.)

2. Use the addition property of equality. (Frames 3-10)

3. Simplify equations, and then use the addition property of equality. (Frames 11-18)

4. Use the multiplication property of equality. (Frames 19-26)

5. Simplify equations, and then use the multiplication property of equality. (Frames 27-33)

---

1. A _____ equation has the form $ax + b = 0$. | linear

2. Is the equation $8x + 4 = 0$ linear? (*yes/no*) | yes

3. Solve the equation $x - 3 = 14$. To make the left side simply x, add ____ to both sides of the equation. | 3

    $$x - 3 = 14$$
    $$x - 3 + \_\_\_\_ = 14 + \_\_\_\_$$ | 3; 3
    $$x = \_\_\_\_$$ | 17

    The solution is ____. Check by substituting this in $x - 3 = 14$. | 17

    $$\_\_\_\_ - 3 = 14$$ | 17
    $$\_\_\_\_ = 14 \quad (\textit{true/false})$$ | 14; true

4. Solve the equation $x - 1 = 5$ by the addition property of equality. You can add the same _____ | number
    to (*one/both*) sides(s) of the _____. What is | both; equation
    the additive inverse of -1? ____ Add ____ to | 1; 1
    _____ sides of the equation. | both

    $$x - 1 + \_\_\_\_ = 5 + \_\_\_\_$$ | 1; 1

    The result is $x = \_\_\_\_$. The solution of | 6
    $x - 1 = 5$ is ____. | 6

2.2 The Addition and Multiplication Properties of Equality

5. Instead of adding the same quantity to both sides of an equation, we may also _____ the same quantity from both sides.

   subtract

6. To solve $y + 9 = 2$, subtract ____ from both sides.

   $y + 9 - \_\_\_\_ = 2 - \_\_\_\_$

   $y = \_\_\_\_$

   9

   9; 9

   −7

In Frames 7–10, decide which quantity to add or subtract on both sides of each equation. Then add or subtract that quantity, and solve the equation for the given variable.

7. $m + 4 = 16$

   Subtract ____. $m + 4 - \_\_\_\_ = _____$

   $m = \_\_\_\_$

   4; 4; 16 − 4

   12

8. $x - 17 = -10$

   Add ____.

   $x _____ = -10 + _____$

   $x = \_\_\_\_$

   17

   − 17 + 17; 17

   7

9. $8 + p = -3$

   Subtract ____. $_____ = _____$

   $p = \_\_\_\_$

   8; 8 + p − 8; −3 − 8

   −11

10. $-2 + k = 44$

    Add ____.

    $-2 + k + _____ = _____$

    $k = \_\_\_\_$

    2

    2; 44 + 2

    46

11. Solve the equation $x + 13 - 2 = 15$. First ____ terms.

    $x + \_\_\_\_ = 15$

    $x + 11 - _____ = 15 _____$

    $x = \_\_\_\_$

    combine

    11

    11; − 11

    4

Chapter 2 Solving Equations and Inequalities

The solution of the equation x + 13 - 2 = 15 is
_____. Check this.

	4
	Substitute 4 for x:
	4 + 13 - 2 = 15
	15 = 15; true
	The solution is 4.

12. Solve the equation 8x + 3 - 7x + 2 + 1 = 9.
First add or subtract where possible to _____   combine
terms.

$$8x + 3 - 7x + 2 + 1 = 9$$
$$(\qquad) + (\qquad) = 9$$
$$\underline{\qquad} = 9$$

Finish the solution.

	8x - 7x; 3 + 2 + 1
	x + 6
	x + 6 - 6 = 9 - 6
	x = 3

You subtracted the same quantity from _____ sides   both
of the equation by applying the _____ prop-   addition
erty of equality. The resulting equation is
_____. The solution is _____. Check this.   x = 3; 3

13. Solve -8 + 9k + 5k - 2 - 13k = 5

$$(\qquad) + (\qquad) = 5$$
$$\underline{\qquad} = 5$$
$$k - 10 + \underline{\quad} = 5 + \underline{\quad}$$
$$k = \underline{\quad}$$

The solution is _____. Check this.

	9k + 5k - 13k; -8 - 2
	k - 10
	10; 10
	15
	15

14. Solve -6 = 9r - 5r + 2 - 3r.

	-6 = r + 2
	-6 - 2 = r + 2 - 2
	-8 = r

The solution is _____. Check this.   -8

2.2 The Addition and Multiplication Properties of Equality

15. Solve $5p + 9 = 4p + 3$. Transpose the variable terms to the left side by _____ 4p from both sides. | subtracting

$$5p + 9 = 4p + 3$$
$$5p + 9 - 4p = 4p + 3 - \underline{}$$ | 4p
$$\underline{} = \underline{}$$ | p + 9; 3
$$p + 9 - 9 = 3 - \underline{}$$ | 9
$$\underline{} = \underline{}$$ | p; −6

The solution is _____. Check this. | −6

16. Solve $3(2x + 7) - (8 + 5x) = 16$. First, use the _____ property to simplify the equation. | distributive

$$3(2x + 7) - (8 + 5x) = 16$$
$$\underline{} = 16$$ | 6x + 21 − 8 − 5x
$$\underline{} = 16$$ | x + 13
$$x + 13 - \underline{} = \underline{}$$ | 13; 16 − 13
$$x = \underline{}$$ | 3

The solution is _____. Check this. | 3

Use the distributive property to solve the following equations. Check each solution.

17.
$$-6(2x - 4) = -11x + 7$$
$$\underline{} = -11x + 7$$ | −12x + 24
$$-12x + 24 + \underline{} = -11x + 7 + \underline{}$$ | 12x; 12x
$$\underline{} = \underline{} + 7$$ | 24; x

Subtract _____ from both sides. | 7
$$\underline{} = \underline{}$$ | 24 − 7; x + 7 − 7
$$\underline{} = x$$ | 17

The solution is _____. | 17

Chapter 2 Solving Equations and Inequalities

18. $9x - 17 = 2(5x - 6)$

   $9x - 17 = $ _____ $10x - 12$

   $9x - 17 - 9x = 10x - 12 - 9x$

   $-17 = x - 12$

   $-17 + 13 = x - 12 + 12$

   $-5 = x$

   The solution is _____. -5

19. The multiplication property of equality says
   that _____ sides of an equation can be mul-   both
   tiplied by the _____ quantity (except zero). same
   For example, to solve the equation $5x = 30$,
   multiply both sides by _____, which is the $1/5$
   _____ of 5.   reciprocal

   $5x = 30$

   $(\quad)5x = (\quad)30$ $\frac{1}{5}; \frac{1}{5}$

   $(\quad \cdot 5)x = \_\_\_ \cdot 30$ $\frac{1}{5}; \frac{1}{5}$

   $(\quad)x = \_\_\_$ $1; 6$

   $\_\_\_\_ = 6$ x

   Is x alone on the left? (yes/no)   yes
   The solution to the equation $5x = 30$ is _____. 6

20. To solve $\frac{1}{3}x = 4$, multiply both sides by 3 because
   3 is the _____ of $1/3$.   reciprocal

   $\frac{1}{3}x = 4$

   $(\quad)\frac{1}{3}x = (\quad)4$ 3; 3

   $(3 \cdot \frac{1}{3})x = \_\_\_$ 12

   $(\quad)x = 12$ 1

   $x = 12$

21. Instead of multiplying both sides of an equation
   by the same quantity, we can also _____ both divide
   sides by the same nonzero quantity.

2.2 The Addition and Multiplication Properties of Equality

22. For example, to solve $-3x = 27$, we can _____ both sides by _____, the coefficient of the _____.

    $$-3x = 27$$
    $$\frac{-3x}{-3} = \underline{}$$

    Since $\frac{-3}{-3} =$ _____,

    $$x = \underline{}.$$

    Check this solution by replacing x with _____ in the original equation.

    | divide |
    | -3 |
    | variable |
    | |
    | |
    | $\frac{27}{-3}$ |
    | |
    | 1 |
    | -9 |
    | |
    | -9 |

23. Solve the equation $\frac{x}{5} = 3$. The expression $\frac{x}{5}$ is the same as ( )x, or $\frac{1}{5}x$. The reciprocal of $\frac{1}{5}$ is _____.

    $$\frac{x}{5} = 3$$
    $$()\frac{x}{5} = ()3$$
    $$(\cdot \frac{1}{5})x = 15$$
    $$\underline{} = 15$$

    The solution is _____. Check this solution.

    | $\frac{1}{5}$ |
    | 5 |
    | |
    | |
    | 5; 5 |
    | 5 |
    | 1x or x |
    | 15 |

24. Solve $\frac{x}{2} = 3$. The expression $\frac{x}{2}$ is the same as _____. Multiply both sides by the reciprocal of _____, which is _____.

    $$\frac{x}{2} = 3$$
    $$()\frac{x}{2} = ()3$$
    $$\underline{} = \underline{}$$

    The solution is _____. Check this solution.

    | $\frac{1}{2}x$ |
    | 1/2; 2 |
    | |
    | 2; 2 |
    | x; 6 |
    | 6 |

Chapter 2 Solving Equations and Inequalities

25. Solve $\frac{3}{4}x = 9$. Multiply both sides by the reciprocal of _____, which is _____. 3/4; 4/3

$$\frac{3}{4}x = 9$$
$$(\quad)\frac{3}{4}x = (\quad)9$$ $\frac{4}{3}$; $\frac{4}{3}$
$$\underline{\quad} = \underline{\quad}$$ x; 12

The solution is _____. Check this solution.   12

26. Solve $\frac{5x}{8} = 10$. The expression $\frac{5x}{8}$ is the same as _____. Multiply both sides of the equation by the reciprocal of _____, which is _____. $\frac{5}{8}x$; 5/8; 8/5

$$\frac{5x}{8} = 10$$
$$(\quad)\frac{5x}{8} = (\quad)10$$ $\frac{8}{5}$; $\frac{8}{5}$
$$\underline{\quad} = \underline{\quad}$$ x; 16

The solution is _____. Check this solution.   16

27. To solve $8k + 7k = 60$, first add 8k and 7k to get _____. The equation is then _____ The solution is _____. Check this solution.   15k; 15k = 60; 4

Solve each equation. Check all solutions.

28. $12z - 5z = -63$ The solution is _____.   -9

29. $15k - 9k = -18$ The solution is _____.   -3

30. $12p - 11p = 0$ The solution is _____.   0

31. Solve $-x = 4$. This equation gives the value of $-x$, but you want the value of _____. To find this, multiply both sides by _____.   x; -1

2.3 More on Solving Linear Equations

( )(−x) = ( )4	−1; −1
( )(−1)x = _____	−1; −4
( )x = _____	1; −4
_____ = −4	x
The solution is _____. Check this solution.	−4

32. Solve −m = −15.   The solution is _____. | 15

33. Solve the equation $3(6x - 2) - (5 + 19x) = 2$.
    First, use the distributive property.

    $3(6x - 2) - (5 + 19x) = 2$

    _____ − = 2 | $18x - 6 - 5 - 19x$

    Then combine like terms.

    _____ = 2 | $-x - 11$

    Add _____ to both sides. | 11

    _____ = _____ | $-x - 11 + 11$; $2 + 11$
    $-x =$ _____ | 13

    Multiply both sides by _____. | −1

    _____ (−x) = _____ ( ) | −1; −1(13)
    x = _____ | −13

    The solution is _____. | −13

2.3 More on Solving Linear Equations

[1] Learn the four steps for solving a linear equation and how to use them. (See Frames 1–10 below.)

[2] Solve equations with fractions as coefficients. (Frames 11–12)

[3] Solve equations with decimals as coefficients. (Frames 13)

[4] Recognize equations with no solutions or infinitely many solutions. (Frames 14–15)

[5] Write a sentence or sentences using an algebraic expression. (Frames 16–24)

Chapter 2 Solving Equations and Inequalities

1. There are four steps that may be necessary to solve a _____ equation.   **linear**

    (1) Simplify each side of the equations by _____ like terms.   **combining**

    (2) Use the _____ property to get all   **addition**
    _____ on one side, and all   **variables**
    _____ on one side.   **numbers**

    (3) Use the _____ property to get a   **multiplication**
    final equation of the form ____ = a number.   **x**

    (4) Check the _____.   **solution**

2. Solve the equation 3x − 5x + 6 = 22.

    Step 1 Simplify.   (   )x + 6 = 22   **(3 − 5)**
                                                _____ + 6 = 22   **−2x**

    Step 2 Use the   −2x + 6 − ____ = 22 − _____   **6; 6**
    addition property.
                                                   −2x = ____   **16**

    Step 3 Use the multi- $\dfrac{-2x}{\_\_} = \dfrac{16}{\_\_}$ **−2; −2**
    plication property.
    (Divide both
    sides by −2.)
                                                   x = ____   **−8**

    Step 4 Check   3(   ) − 5(   ) + 6 = 22   **−8; −8**
    the solution.
                                     _____ + _____ + 6 = 22   **−24; 40**
                                                   _____ = 22   **22**

    Does the solution check? *(yes/no)*   **yes**

3. Solve the equation 5k + 2 − 8k = 17.

    Step 1 Simplify.   (   ) + 2 = 17   **5k − 8k**
                                             _____ = 17   **−3k + 2**

    Step 2 Addition   −3k + 2 = 17
    property.
                                     −3k + 2 ____ = 17 ____   **− 2; − 2**

    Step 3 Multiplication $\dfrac{-3k}{\_\_} = \dfrac{15}{\_\_}$ **−3; −3**
    property.
                                             k = ____   **−5**

2.3 More on Solving Linear Equations

Step 4 Check the solution. $\quad 5() + 2 - 8() = 17$	$-5;\ -5$
$\underline{} + 2 + \underline{} = 17$	$-25;\ 40$
$\underline{} = 17$	17
Does the solution check? *(yes/no)*	yes

4. Solve the equation $2(m - 5) + 1 = -19$ distributive property to eliminate parentheses.

Step 1 $\quad 2(m - 5) + 1 = -19$	
$\underline{} - \underline{} + 1 = -19$	$2m - 10$
$\underline{} = -19$	$2m - 9$
Step 2 $\quad 2m - 9 + \underline{} = -19 + \underline{}$	$9;\ 9$
$\underline{} = \underline{}$	$2m;\ -10$
Step 3 $\quad \dfrac{2m}{\underline{}} = \dfrac{-10}{\underline{}}$	$2;\ 2$
$m = \underline{}$	-5
The solution is $\underline{}$.	-5
Step 4 Does the solution check in the original equation? *(yes/no)*	yes

5. Solve $3(2a - 3) - 8 = 2a + 3$.

Step 1 $\quad \underline{} - 8 = 2a + 3$	$6a - 9$
$\underline{} = 2a + 3$	$6a - 17$
Step 2 $\quad \underline{} = 3$	$4a - 17$
$4a = \underline{}$	$3 + 17$ (or 20)
Step 3 $\quad a = \underline{}$	5
The solution is $\underline{}$.	5
Step 4 Does the solution check in the original equation? *(yes/no)*	yes

6. Solve $5(4x + 3) = -3(x + 5) - 16$.

$\underline{} = \underline{} - 16$	$20x + 15;\ -3x - 15$
$20x + 15 = \underline{}$	$-3x - 31$
$\underline{} = -31$	$23x + 15$

Chapter 2 Solving Equations and Inequalities

23x = _____	−46
x = _____	−2
The solution is _____.	−2

Step 4 Does the solution check in the original equation? (yes/no)   yes

7. −3(r − 5) + 2r − 6 = r + 1

　　　　　　　　The solution is _____.   4

8. 2(3 − 2x) + 2 = x − 2

　　　　　　　　The solution is _____.   2

9. 3(4 − 2k) + 6k = −2(k − 1)

12 − 6k + 6k = −2k + 2
12 = −2k + 2
10 = −2k
−5 = k

　　　　　　　　The solution is _____.   −5

10. −(k − 6) + 2(k + 8) = 4k + 4

　　　　　　　　The solution is _____.   6

2.3 More on Solving Linear Equations

11. Solve the equation $\frac{3}{4}x - \frac{1}{3}x = \frac{1}{2}x - 1$. The least _____ denominator of all the fractions is ____. _____ both sides by ____ to find an equivalent equation with only integers as coefficients.

	common; 12
	Multiply; 12

    $$\frac{3}{4}x - \frac{1}{3}x = \frac{5}{6}x - 5$$

    Multiply both sides by ____.   12

    $(\ )(\frac{3}{4}x - \frac{1}{3}x) = (\ )(\frac{5}{6}x - 5)$ 12; 12

    $(\ )(\frac{3}{4}x) - (\ )(\frac{1}{3}x) = (\ )(\frac{5}{6}x) - (\ )(5)$ 12; 12; 12; 12

    ___ x - ___ x = ___ x - ___   9; 4; 10; 60

    Step 1   ___ x = ___ x - ___   5; 10; 60

    Step 2   5x - ___ x = ___   10; -60

               ___ x = ___   -5; -60

    Step 3 $\frac{-5x}{\_\_\_} = \frac{-60}{\_\_\_}$ -5; -5

               x = ___   12

    The solution is ___.   12

    Step 3 Does the solution check? (yes/no)   yes

12. Solve the equation $\frac{1}{5}x - 2 = \frac{2}{3}x - \frac{2}{5}x$. The least common denominator is ____.   15

    $$\frac{1}{5}x - 2 = \frac{2}{3}x - \frac{2}{5}x$$

    Multiply both sides of the equation by ____.   15
    The equivalent equation with only integers as coefficients is _____.   3x - 30 = 10x - 6x

    The solution is ____.   -30

13. To solve the equation $.4y + .6(50 - y) = .4(68)$, multiply both sides by ____. To do this you can move the decimal places in the equation ____ place(s) to the right. This gives the equation _____.

	10
	1
	4y + 6(50 - y) = 4(68)

64   Chapter 2   Solving Equations and Inequalities

Solve the equation using the four steps.

The solution is ____.	14

14. It is possible for a linear equation to have one solution, ____ solution, or ____ many solutions. | no; infinitely

15. Solve the equation $x + 2(x - 1) = 3x + 5$.

$$x + 2(x - 1) = 3x + 5$$
$$x + \underline{} - \underline{} = 3x + 5 \qquad \text{2x; 2}$$
$$\underline{} - \underline{} = 3x + 5 \qquad \text{3x; 2}$$

Subtract ____ from both sides. | 3x

$$3x - 2 - \underline{} = 3x + 5 - \underline{} \qquad \text{3x; 3x}$$
$$- 2 = 5 \quad (true/false) \qquad \text{false}$$

The resulting ____ statement means that the equation has ____ solution. | false; no

16. Solve the equation $-3x + 18 = -3(x - 6)$.

$$-3x + 18 = -3(x - 6)$$
$$-3x + 18 = \underline{} \qquad -3x + 18$$

Add ____ to both sides. | 3x

$$-3x + 18 + \underline{} = \underline{} \qquad \text{3x; } -3x + 18 + 3x$$
$$18 = 18 \quad (true/false) \qquad \text{true}$$

The statement is ____, but the variable has disappeared. ____ real number is a solution. The solution should be indicated as "____ real numbers." | true; Any; all

2.3 More on Solving Linear Equations

17. Two numbers have a sum of 15. One of the numbers is 7. You find the other number by subtracting: _____ − ___ . The other number is ___ .

    15 − 7; 8; 8

18. Two numbers have a sum of 34. One number is y. Write an algebraic expression for the other number: _____ .

    $34 - y$

19. The product of two numbers is 126. One number is 14. Find the other number by _____ . The other number is _____ = ___ .

    dividing

    $\frac{126}{14}$; 9

20. The product of two numbers is 92. One number is r. Write an algebraic expression for the other number: _____

    $\frac{92}{r}$

21. Cliff bought t tapes. He gave 3 tapes to his brother. How many did he have left? Write the answer as an algebraic expression. _____

    $t - 3$

22. The Student Council has five times as many members as it has officers. If the Council has q officers write an algebraic expression for the number of members. _____

    $5q$

23. Reggie made h hits in a game and Kevin made 2 more hits than Reggie. Write an algebraic expression for the number of hits Kevin made. _____

    $h + 2$

24. Yolanda has k dimes. Write an algebraic expression for the value of the dimes in cents. _____

    $10k$

66 Chapter 2 Solving Equations and Inequalities

25. A concession stand at the stadium took in m dollars in ten-dollar bills and n dollars in five-dollar bills. How many bills are there in the two denominations? _____

$\dfrac{m}{10} + \dfrac{n}{5}$

2.4 An Introduction to Applications of Linear Equations

[1] Learn the six steps to be used to solve an applied problem. (See Frames 1-2 below.)

[2] Solve problems involving unknown numbers. (Frames 3-7)

[3] Solve problems involving sums of quantities. (Frames 8-12)

[4] Solve problems involving supplementary and complementary angles. (Frames 13-14)

1. In general, there are six steps to help you solve an applied problem.

   Step 1 _____ the problem carefully, and choose a _____ to represent the _____ number. Write down what the _____ represents.

   Read
   variable; unknown
   variable

   Step 2 Write down a mathematical _____ using the variable for any other _____ quantities. _____ any figures or diagrams that apply.

   expression
   unknown
   Draw

   Step 3 Translate the problem into an _____.

   equation

   Step 4 _____ the equation.

   Solve

   Step 5 _____ the question asked.

   Answer

   Step 6 Check your solution in the _____ of the problem.

   words

2.4 An Introduction to Applications of Linear Equations

2. In each of the following lines, one word or expression does not belong with the others. Cross out the item that does not belong.

   Added to   More than   Difference   Increased by

   Sum   Divided by   Quotient   Goes into

   Times   Product   Decreased by   Multiplied by

   Difference   Subtracted from   Less   Divides

Cross out:

Difference

Sum

Decreased by

Divides

Solve the following applied problems using the six steps.

3. The sum of a number and 8 is 12. Find the number.

   Step 1 _____

   Step 2 _____

   Step 3   The sum of   is   12.
            a number and 8
              ↓   ↓   ↓

   Step 4

                    x = _____

   Step 5 The number is ____.

   Step 6 The sum of _____ and 8 is 12.
           The solution (checks/does not check).

Let x = the number.

There are no other unknown quantities.

$x + 8 = 12$

$x + 8 - 8 = 12 - 8$

4

4

4

checks

4. A number minus 3 equals 9. What is the number?

                                      x = _____

   The number is _____.

Let x = the number.
$x - 3 = 9$
$x - 3 + 3 = 9 + 3$
12
12

5. Three times a number is equal to 1 more than twice a number. Find the number.

Let x = the number.

68   Chapter 2   Solving Equations and Inequalities

Three times a number	is equal to	1	more than	twice a number.
↓	↓	↓	↓	↓

3x = 1 + 2x

The number is _____.   1

6. Three times the difference of a number and 8 is 21. Find the number.

Let x = the number.

Three	times	the difference of a number and 8	is	21.
↓	↓	↓	↓	↓

$$3 \cdot (x - 8) = 21$$

3(   )   = 21   x − 8

_____ − _____ = 21   3x − 24

3x = __   45

x = __   15

The number is _____.   15

7. The difference of 4 times a number and 8 is 16. What is the number?

Let x = the number.

	The difference of 4 times a number and 8	is	16.
	↓	↓	↓

4x − 8 = 16

4x = __   24

x = __   6

The number is _____.   6

8. The sum of two numbers is 18. One of the numbers is 2 more than the other. Find the two numbers.

Step 1

Step 2

Let x = the smaller number.
x + 2 = the larger number.

2.4 An Introduction to Applications of Linear Equations

Step 3

The total	is	the smaller	plus	the larger.
↓	↓	↓	↓	↓

$18 = x + x + 2$

$x = $ _____   8

$x + 2 = $ _____   10

One number is ___ and the other number is ____.   8; 10

Check: _____ . $8 + 10 = 18$

9. Smith and Jones both ran for president of the Math Lovers' Club. Smith received 85 more votes than Jones. The total number of votes cast was 309. Find the number of votes received by each person.

   Let x = the number of votes received by Jones. Then Smith received _____ votes. $x + 85$

   The sum of the votes of the two people (*is/is not*) the same as the "total number of votes cast." Write an equation and solve it.   is

   _____

   $x = $ _____ and $x + 85 = $ _____ $x + x + 85 = 309$; 112; 197

   Votes for Jones _____   112

   Votes for Smith _____   197

   Check: _____ + _____ = 309   112; 197; 309

10. A farmer has 7 more than three times as many hens as roosters. The total number of these birds is 327. Find the number of hens she has. Let x = the number of roosters. Then the number of hens is 7 more than three times _____, in   x

    symbols, _____. $3x + 7$

    The words "total number of these birds is 327" tells you a fact about the (*sum/product*) of two numbers. Finish the problem.   sum

    $x + 3x + 7 = 327$
    $4x + 7 = 327$
    $4x = 320$
    $x = 80$
    $3x + 7 = 247$

Chapter 2 Solving Equations and Inequalities

x = _____. | 80
x gives the number of _____. | roosters
The number of hens is _____. | 247

11. Consecutive integers are in a row, such as 5 and
____, or 11 and _____. Consecutive odd integers | 6; 12
are _____ integers in a row, such as 17 and ___, | odd; 19
or −7 and _____. If −10 is the first of three | −5
consecutive even integers, the second and third
are _____ and _____. | −8; −6

12. **If the smaller of two consecutive even integers is tripled, the result is 20 more than twice the larger. Find the two integers.**
Let x = the smaller integer, with _____ for the | x + 2
larger. Complete an equation.

Smaller tripled is 20 more than twice larger
_____   ___   20   +   _____ | 3x; =; 2(x + 2)

Solve the equation to find that x = _____. The | 24
smaller integer is _____, and the larger is | 24
x + 2 = _____. | 26

13. In geometry a unit that measures angles is a ____. | degree
If the sum of the measures of two angles is 90°,
the angles are said to be _____. Two | complementary
angles, the sum of whose measures are _____, are | 180°
said to be supplementary.

14. **Find the measure of an angle whose supplement measures 20° more than 3 times the measure of its complement.**
Let x = _____.

_____ − x = the measure of the complement | the measure of
180 − x = _____. | the angle
 | 90
 | the measure of
 | the supplement

2.5 Formulas and Applications from Geometry   71

Write an equation and solve it.

The supplement	measures	20°	more than	3 times the complement
↓	↓	↓	↓	↓

$180-x=20+3(90-x)$
$180-x=20+270-3x$
$2x=110$
$x=55$

The angle measures _____.   55°

2.5 Formulas and Applications from Geometry

[1] Solve a formula for one variable given the values of the other variables. (See Frames 1–12 below.)

[2] Solve a formula for a specified variable. (Frames 13–16)

[3] Use a formula to solve a geometric application. (Frames 17–19)

[4] Solve problems about angle measures. (Frames 20–23)

---

1. Many applied problems can be solved using _____.   formulas
   Formulas exist for geometric figures, for distance, and for interest earned on bank savings.

See Appendix A in your textbook for a list of formulas. Some of these formulas are illustrated in the following.

2. The figure shown is a _____.   square

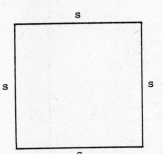

We let s = the length of a _____,   side
A = the _____,   area
_____ = the perimeter.   P
A = _____ s^2
P = _____ $4s$

72   Chapter 2 Solving Equations and Inequalities

3. The figure shown
   is a _____ .   rectangle
   L = the _____   length
   W = the _____   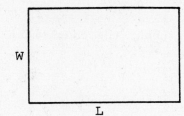   width
   A = _____   LW
   P = _____   2L + 2W

4. This figure is
   a _____ .   triangle
   h = the _____   height
   b = the _____   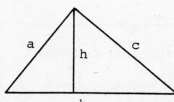   base
   a, c = the lengths of the
          other _____   sides
   A = _____ $\frac{1}{2}bh$
   P = _____   a + b + c

5. This figure is a
   trapezoid.

                          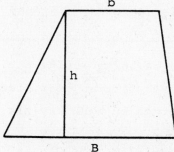

   h = the _____   height
   b = the shorter parallel _____   side
   B = the _____ parallel side   longer
   A = _____ $\frac{1}{2}(b + B)h$

6. Often we are given a formula and the value of
   some of the variables in the formula. For example, suppose we are given the formula

                 $A = \frac{1}{2}bh,$

   and the fact that b = 42 meters and h = 50
   meters. To find ____, replace b with ____   A; 42
   and h with ____.   50

         $A = \frac{1}{2}($ $)($ $) = $ ____ square meters   42; 50; 1050

2.5 Formulas and Applications from Geometry

Find the value of the missing variable in each of the following.

7. $P = 2L + 2W$; $L = 80$, $W = 30$

    $P = 2() + 2()$ 80; 30

    $P = \underline{} + \underline{} = \underline{}$ 160; 60; 220

8. $d = rt$; $d = 50$, $r = 10$

    $() = ()t$ 50; 10

    $\underline{} = t$ 5

9. $P = 2L + 2W$; $P = 900$, $L = 300$

    $\underline{} = 2() + \underline{}$ 900; 300; 2W

    $900 = \underline{} + 2W$ 600

    $\underline{} = 2W$ 300

    $\underline{} = W$ 150

10. $A = \frac{1}{2}(b + B)h$; $b = 15$, $B = 25$, $h = 6$

    $A = \underline{}$ 120

11. $V = \frac{4}{3}\pi r^3$; $r = 9$, $\pi = 3.14$ $V = \underline{}$ 3052.08

12. $A = \frac{1}{2}(b + B)h$; $A = 600$, $B = 70$, $h = 10$

    $b = \underline{}$ 50

13. Sometimes it is necessary to work many problems using the same formula. This can be made easier if you isolate the one variable you need to find. For example, if you are using the formula for the area of a triangle, you might need to find the base. The formula for the area of a triangle is

    $A = \underline{}$. $\frac{1}{2}bh$

Chapter 2 Solving Equations and Inequalities

To find the base, you would need to get b alone on one side of the equals sign, and all other letters on the other side. To do this here, first multiply both sides by 2.

$$(\quad)A = (\quad)\tfrac{1}{2}bh$$
$$\underline{\quad\quad} = \underline{\quad\quad}$$

2; 2
2A; bh

Now divide both sides by h.

$$\frac{2A}{\underline{\quad}} = \frac{bh}{\underline{\quad}}$$

$$\underline{\quad\quad} = b$$

h; h

$$\frac{2A}{h}$$

This process is called _____ for a specified variable.

solving

14. Solve $P = 2L + 2W$ for W. First, subtract _____ from both sides.

2L

$$P - \underline{\quad} = 2L + 2W - \underline{\quad}$$
$$P - 2L = \underline{\quad}$$

2L; 2L
2W

Now divide both sides by _____, and the result is

2

$$\underline{\quad\quad} = W.$$

$$\frac{P - 2L}{2}$$

15. Solve $C = 2\pi r$ for r.

$$r = \underline{\quad\quad}$$

$$\frac{C}{2\pi}$$

16. Solve $A = \tfrac{1}{2}(b + B)$ for h.

$$h = \underline{\quad\quad}$$

$$\frac{2A}{b + B}$$

Solve.

17. The formula for the area of a rectangle is _____. Suppose the area of a particular rectangle is 20. If the length of the rectangle is 10, find the width.

$A = LW$

2.5 Formulas and Applications from Geometry

$A = LW$		
___ = ___ · W	Substitute values	20; 10
___ = W	Divide by 10	2
or W = ___		2

18. The formula for the area of a triangle is _____. $A = \frac{1}{2}bh$
    If the height of the triangle is 20 inches and its
    area is 80 square inches, then the length of the
    base of the triangle is _____.   8 inches

19. The perimeter of a rectangle is 100 feet and the
    width is 10 feet. The length of the rectangle
    is _____.   40 feet

20. The two _____ lines   intersecting
    form angles numbered 1, 2, 3, 4.
    Angles 1 and ___ are opposite   3
    each other. They are called
    _____ angles.   vertical
    Angles ___ and ___ are also   2; 4
    vertical angles. Vertical angles
    have _____ measure.   equal

21. The sum of the measures of angles 1 and 2 in
    in Exercise 20 is _____, which is the same as   180°
    the measure of a _____ angle. Other pairs   straight
    of angles whose measures add to 180° are
    _____, _____, and _____   2 and 3; 3 and 4;
                                                       4 and 1

Find x in each figure. Then find the measure of each marked angle.

22. The sum of the measure
    of two angles is _____. $(4x + 10)°$ $(3x - 5)°$ 180°

76 Chapter 2 Solving Equations and Inequalities

Write and solve an equation.	
_____ + _____ = 180	4x + 10; 3x − 5
x = ____	25
The angle marked (4x + 10)° measures ____. The	110°
angle marked (3x − 5)° measures _____.	70°
23. The marked angles are _____ angles so they have _____ measures. 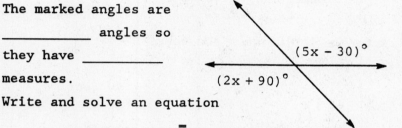 Write and solve an equation _____ = _____	vertical equal 5x − 30; 2x + 90
x = _____	40
The angle marked (5x − 30)° measures _____. The	170°
angle marked (2x + 90)° also measures _____.	170°

2.6 Ratios and Proportions

[1] Write ratios. (See Frames 1–4 below.)

[2] Decide whether proportions are true. (Frames 5–10)

[3] Solve proportions. (Frames 11–14)

[4] Solve applied problems using proportions. (Frames 15–18)

[5] Solve problems involving unit pricing. (Frames 19–20)

---

1. Ratios are used to _____ two numbers or quantities. The ratio of the number a to the number b is written as _____ or _____.	compare a:b; $\frac{a}{b}$

2.6 Ratios and Proportions

Write each ratio.

2. 7 men to 9 men _____

$\frac{7}{9}$

3. 6 weeks to 28 days
Since the ratio is used in comparing units of measure, the _____ must be the same before a ratio can be written. Since

units

    1 week = _____ days,

7

    6 weeks is _____ days, and the ratio is

42

    _____ = _____ (in lowest terms).

$\frac{42}{28}$; $\frac{3}{2}$

4. 5 quarts to 11 pints
Since 1 quart = _____ pints, 5 quarts is _____ pints. The ratio is _____.

2; 10

$\frac{10}{11}$

5. A ratio compares two numbers. A proportion says that two _____ are equal.

ratios

6. The statement $\frac{a}{b} = \frac{c}{d}$ is a _____.

proportion

7. For a proportion to be true, the _____ must be _____. In Frame 6 the cross products are _____ and _____.

cross products

equal

ad; bc

Decide if the following proportions are true.

8. $\frac{5}{6} = \frac{15}{18}$

    One cross product is ____ · ____ = ____.

5; 18; 90

    Another cross product is ____ · ____ = ____.

6; 15; 90

    Are the cross products equal? (*yes/no*)

yes

    Is the proportion true? (*yes/no*)

yes

Chapter 2 Solving Equations and Inequalities

9. $\dfrac{12}{7} = \dfrac{38}{21}$

   One cross product = _____. | $12 \cdot 21 = 252$
   Other cross product = _____. | $7 \cdot 38 = 266$
   Is the proportion true? (yes/no) | no

10. $\dfrac{29}{5} = \dfrac{145}{25}$ \qquad True? _____ | yes

11. A proportion uses _____ numbers. If _____ of them are known, the fourth can be found. | four; three

Find the value of the missing numbers in Frames 12–14.

12. $\dfrac{x}{9} = \dfrac{12}{54}$

    The cross products must be equal, or $54x =$ _____. | $9 \cdot 12$, or 108
    Divide both sides by 54 to get $x =$ _____. | 2

13. $\dfrac{7}{6} = \dfrac{p}{9}$

    Find the two cross products, and set them equal.

    $7 \cdot 9 =$ ____ | $6p$
    _____ = $6p$ | 63
    _____ = p | $\dfrac{21}{2}$

14. $\dfrac{12}{y} = \dfrac{5}{8}$ \qquad $y =$ _____ | $\dfrac{96}{5}$

15. A car goes 253 miles on 11 gallons of gas. How far would it go on 17 gallons of gas?

    Write a proportion. Let $x =$ the unknown number of miles.

    $\begin{array}{r}\text{miles} \rightarrow \\ \text{gallons} \rightarrow\end{array} \dfrac{253}{\_\_\_\_} = \dfrac{x}{17} \begin{array}{l}\leftarrow \text{unknown miles} \\ \leftarrow \text{gallons}\end{array}$ | 11

    Multiply. \qquad $11x =$ _____ | 4301
    \qquad\qquad $x =$ _____ miles | 391

2.6 Ratios and Proportions

16. **Nine pills cost $2.25. Find the cost of 20 pills.**

    Let x = the cost of 20 pills. Write a proportion.

    $$\frac{2.25}{\rule{1cm}{0.4pt}} = \frac{x}{\rule{1cm}{0.4pt}}$$

    9x = _____

    x = _____

    9; 20

    45.00

    $5.00

17. **On a map, 12 centimeters represents 500 kilometers. How many centimeters would be needed for 2000 kilometers?** _____

    48

18. **If 6 nights in a motel cost $180, find the cost for 5 nights.** _____

    $150

19. Many shoppers compare the _____ prices of items of different sizes to decide which size is the best to buy. To do this, _____ the price of each item by the number of units of measure in the item. Then compare the _____ of each unit in the different sizes.

    unit

    divide

    price

20. The regular price of an 18-ounce package of Kitty Nums is $1.49 and the price of a 56-ounce bag is $3.49. The 18-ounce package is on sale for $.99. Which of the three is the best buy? Find the unit price to three decimal places.

    18-ounce (regular) $\frac{\$1.49}{\rule{1cm}{0.4pt}}$ = _____

    56-ounce _____ = _____

    18-ounce (sale) _____ = _____

    The best buy is _____.

    18; $.083

    $\frac{3.49}{56}$; .062

    $\frac{\$.99}{18}$; $.055

    18-ounce (sale)

80   Chapter 2   Solving Equations and Inequalities

2.7 Applications of Percent: Mixture, Interest, and Money

[1] Learn how to use percent in problems involving rates. (See Frames 1-9 below.)

[2] Learn how to solve problems involving mixtures. (Frames 10-11)

[3] Learn how to solve problems involving simple interest. (Frames 12-15)

[4] Learn how to solve problems involving denominations of money. (Frame 16)

---

1. A percent, like a fraction or decimal, represents some part of a whole. When we speak of percents we are referring to hundredths or parts of a hundred. Something has been divided into one hundred parts and we are talking about some of them. For example, 25% means that something has been divided into one _____ parts and we are talking about ____ of them. Also, 50% means that something has been divided into one _____ parts and we are talking about ____ of them, and 100% means that something has been divided into one _____ parts and we are talking about all _____ of them.

   hundred
   25

   hundred
   50

   hundred
   100

2. Quite often a fraction or a decimal must be expressed as a percent. If we begin with a fraction, we must first convert the fraction to a decimal. To express the decimal as a percent, the decimal point is moved two places to the right and a % sign is attached. For example, to express the fraction 3/4 as a percent, first express 3/4 in decimal form as _____. Then move the decimal point ____ places to the _____ and attach a % sign, giving _____ as a result.

   .75
   two; right
   75%

2.7 Applications of Percent: Mixture, Interest, and Money

Common fraction	Decimal fraction	Percent
3/4	.75	75% ← Attach percent sign

3. Percents may be changed to decimals by reversing the above process. For example, 50% becomes .5 and 100% becomes 1. The decimal point was moved two places to the left and the percent sign (%) dropped.

Change decimals to percents:	Change percents to decimals:	
.8 = _____	25% = _____	80%; .25
.75 = _____	142% = _____	75%; 1.42
.007 = _____	87% = _____	.7%; .87
1.4 = _____	85.7% = _____	140%; .857
3.017 = _____	5% = _____	301.7%; .05

4. To find 5% of 780, 5% (rate) will be multiplied by 780. Before a rate can be multiplied it must be changed to a _____.   decimal

$$.05 \times 780 \text{ or } \begin{array}{r} 780 \\ \times\ .05 \\ \hline 39.00 \end{array} \text{ Therefore, 5\% of 780 is \_\_\_.}$$

    39

5. Solve for percentage in the following problems.

12% of 100 _____	40% of 380 _____	12; 152
15% of 60 _____	12 1/2% of 1048 _____	9; 131
6% of 732 _____	37.5% of 510 _____	43.92; 191.25
25% of 800 _____	8% of 120 _____	200; 9.6
20% of 34 _____	115% of 9280 _____	6.8; 10,672

6. Tinker Machine Corporation paid its employees 8% of total sales in the form of profit sharing. If total sales were $275,000 what amount was shared with employees? _____ $22,000

82   Chapter 2 Solving Equations and Inequalities

7. The rate for property tax purposes is 2.5% of cash value. How must would the owner of a home valued at $85,000 pay in taxes? _____ $2125

8. How many grams of pure acetic acid are contained in 500 grams of a 35% acetic acid solution?

   _____   175 grams

9. A bank teller has 43 quarters in a drawer. How much are the quarters worth? _____ $10.75

10. A chemist needs to mix 40 liters of a 10% solution with some 50% solution to get a mixture which is 40% strong. How many liters of the 50% solution should be used?

    Let x be the number of liters of the 50% solution. The amount of pure chemical in this solution is the _____ of the number of liters of solution and the percent strength, or   product

    liters of pure in 50% = _____ .   .50x

    The number of liters of pure in the 40 liters of 10% solution is

    liters of pure in 10% = _____ .   40(.10) = 4

    Fill in the following box diagram to summarize the information.

    | Number of liters of solution | | 40 | | + | ___ | | = | ___ | | x; x + 40 |
    | Rate of concentration | | ___ | | | .50 | | | .40 | | .10 |

2.7 Applications of Percent: Mixture, Interest, and Money

The amount of pure chemical before and after mixing must be equal, so

    pure in 10% + pure in 50% = pure in 40%

    _____ + _____ = _____ .   4; .50x; .40(x + 40)

Solve the equation.

    4 + .50x = _____   .40x + 16

Subtract 4 from both sides.

    .50x = .40x + _____   12

    _____ = 12   .10x

    x = _____   120

A total of _____ liters of 50% solution must be used.   120

11. How many gallons of 40% antifreeze must be mixed with 50 gallons of a 70% solution to get a mixture which is 50% antifreeze?

    Fill in the box diagram.

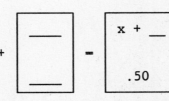

Number of gallons of solution   50; 50

Rate of concentration   .40; .70

Solve the equation

    _____ + _____ = _____ .   .40x; 35; .50(x + 30)

    x = _____   100

    Answer: _____   100 gallons

84   Chapter 2 Solving Equations and Inequalities

12. Simple interest is found by the formula

Interest = (principal)(rate)(time), or I = ____ | prt

where the rate is a percent per year and where time is measured in years or fractions of a year. For example, a person borrowing $2000 for 1 year at 7% interest would pay

Interest = $2000 · ( ) · ____ = ____. | .07; 1; $140

13. Find the simple interest on the given principal, rate, and time.

$1000 at 8% for 2 years   _____ $160
$1500 at 7% for 3 years   _____ $315
$1200 at 6% for 10 months _____ $60

14. George deposits $4000 in a bank paying 6% interest. He leaves the money in for 1/2 year. How much interest does he earn?

The formula is _____. Here p = _____, | I = prt; 4000

r = ____, and t = ____. Thus, | .06; $\frac{1}{2}$

I = ( ) · ( ) · ( ) | 4000; .06; $\frac{1}{2}$
I = ____. | $120

15. A sum of money is invested in two ways. Part is invested at 10%, with $20,000 more invested at 12%. The total annual interest from the two investments is $9000. Find the amount invested at each rate.

Let x be the amount invested at 10%. Then the amount at 12% is _____. The interest at 10% for one year is | x + 20,000

interest at 10% = (principal) × ( ) × ( ) | rate; time
= (x)( )(1) | 10% or .10
= ____. | .10x

2.7 Applications of Percent: Mixture, Interest, and Money

The interest at 12% for one year is

$$(\underline{})(.12)(1) = \underline{}.$$

	x + 20,000; .12x + 2400

The total interest is _____, so

    interest at 10% + interest at 12% = total

$$\underline{} = \underline{} = 9000.$$

	$9000
	.10x; .12x + 2400

Simplify.

$$\underline{} + 2400 = 9000$$
$$.22x = \underline{}$$
$$x = \underline{}$$

	.22x
	6600
	30,000

A total of _____ is invested at 10%, with _____ at 12%.

	$30,000
	$50,000

16. Walt must break into his piggy bank to be able to go to the store. He finds that his bank contains only dimes and nickels. There is a total of 150 coins, with a total value of $11.00. Find the number of each type of coin he has.

Let x = number of nickels. Since the total number of coins is ____, then _____ = number of dimes. Each nickel is worth 5¢, so that the total value of his x nickels is ___. Also, the total value of 150 - x dimes is _____. The value of all the money is $11.00 or _____ cents. Thus,

$$\underline{} + \underline{} = 1100.$$

	150; 150 - x
	5x
	10(150 - x)
	1100
	5x; 10(150 - x)

Solve this equation.

                Nickels: _____
                   Dimes: _____

	80
	70

Chapter 2 Solving Equations and Inequalities

2.8 More About Problem Solving

[1] Use the formula d = rt to solve problems. (See Frames 1-2 below.)

[2] Solve problems involving distance, rate, and time. (Frames 3-4)

[3] Solve problems about geometric figures. (Frames 5-8)

---

1. The basic formula that relates distance, rate, and time is

   _____

   where _____ = distance, _____ = rate, and and ___ = time.

	$d = rt$
	d; r
	t

   Solving for r, we get the form _____. $r = \dfrac{d}{t}$

   Solving for t, we get the form _____. $t = \dfrac{d}{r}$

2. Gale can drive his Honda 120 miles in 5 hours when driving on bad mountain roads. Find his speed, in miles per hour.

   The form of the distance formula needed to solve this problem is

   _____. $r = \dfrac{d}{t}$

   Substitute.

   $$r = \dfrac{\underline{}}{\underline{}}$$ $\dfrac{120}{5}$

   $ = \underline{}$ 24

   Answer: _____   24 miles per hour

3. John can push a baby stroller at 5 miles per hour, and Harriet can pull a red wagon at 4 miles per hour. If they start at the same point, and travel in opposite directions, how long will it take them to be 27 miles apart?

2.8 More About Problem Solving

Complete the following chart. Let x = number of hours until they are 27 miles apart. (Complete the r and t columns first.)

	d	r	t
Harriet			
John			

4x; 4; x

5x; 5; x

Since the total distance between John and Harriet is to be _____ miles,

_____ + _____ = 27

x = _____ hours.

27

5x; 4x

3

4. The distance from Chicago to Toledo is 240 miles. Colette leaves Chicago traveling toward Toledo at 55 miles per hour at the same time that Jed leaves Toledo traveling toward Chicago at 65 miles per hour. How long will it take them to meet?

Let x = the time until they meet. This is the time that each will travel. Complete the chart.

	d	r	t
Colette		55	x
Jed		65	x

55x

65x

Since the total distance that Colette and Jed travel is _____,

_____ + _____ = _____.

x = _____

It will take Colette and Jed ___ hours until they meet.

240 miles

55x; 65x; 240

2

2

Chapter 2 Solving Equations and Inequalities

5. Suppose a rectangle has a width of 9 and a length of 12. Find the perimeter.

   The formula for the perimeter of a rectangle is perimeter = _____. Here W = ___ and L = ___, so that   2L + 2W; 9; 12

   $$\text{perimeter} = 2() + 2()$$ 12; 9
   $$= \underline{} + \underline{}$$ 24; 18
   $$= \underline{}.$$ 42

6. The length of a rectangle is 7 meters more than the width. The perimeter is 50 meters. Find the length and width of the rectangle.

   Write the formula for the perimeter of a rectangle.

   $$P = \underline{}$$ 2L + 2W

   Let x represent the width of the _____.   rectangle
   Since the length is ___ meters more than the   7
   ___, the length is given by ___. In   width; x + 7
   the formula P = 2L + 2W, replace W with ___,   x
   L with ___, and P with ___.   x + 7; 50

   $$\underline{} = 2() + 2()$$ 50; x + 7; x

   Simplify.

   $$50 = \underline{} + \underline{}$$ 2x + 14; 2x
   $$50 = \underline{}$$ 4x + 14

   Subtract ___ from both sides.   14

   $$\underline{} = 4x$$ 36

   Find x.   x = ___   9

   The width is ___ and the length is   9

   ___ + 7 = ___.   9; 16

2.9 The Addition and Multiplication Properties of Inequality

7. Suppose the length of a rectangle is 3 more than the width. The perimeter is 42. Find the length and width.

Let x = width. The _____ = length. The formula for the perimeter of a rectangle is P = _____.	x + 3 2L + 2W
Using our variables, we get $\quad$ P = 2L + 2W $\quad$ ____ = 2( ) + 2( ).	42; x + 3; x
Solve this equation. $\quad$ Width: x = ____ $\quad$ Length: x + 3 = ____	9 12

8. The perimeter of a rectangle is 5 times the width, and the length is 4 more than the width. Find the width and length.

Let x = width of the rectangle. Then _____ = length and ____ = perimeter.	x + 4 5x
Use the formula for the perimeter of a rectangle. $\quad$ P = 2L + 2W $\quad$ _____ = 2( ) + 2( )	5x; x + 4; x
Solve this equation. $\quad$ Width: x = ____ $\quad$ Length: x + 4 = ____	8 12

2.9 The Addition and Multiplication Properties of Inequality

[1] Graph intervals on a number line. (See Frames 1-7 below.)

[2] Use the addition property of inequality. (Frames 8-14)

[3] Use the multiplication property of inequality. (Frames 15-20)

[4] Solve linear inequalities. (Frames 21-26)

[5] Solve applied problems by using inequalities. (Frames 27-35)

[6] Solve three-part inequalities. (Frames 36-43)

90 Chapter 2 Solving Equations and Inequalities

1. Write the meaning of each symbol.

   < _____   is less than

   > _____   is greater than

   ≤ _____   is less than or equal to

   ≥ _____   is greater than or equal to

2. The statement x ≥ −2 tells us to use all real numbers that are _____ than or _____ to −2. To graph these numbers, put a heavy dot at ____ and an arrow to the _____. The heavy dot shows that −2 (*is/is not*) part of the graph and the arrow to the right tells us that we want all numbers _____ than −2. Complete the graph.

   greater; equal

   −2; right

   is

   greater

   x ≥ −2

3. The statement x < 4 tells us to use all real numbers that are _____ than _____. Is 4 itself part of the graph? (*yes/no*). To show that 4 is not part of the graph, use an _____ circle at 4. Complete the graph.

   less; 4

   no

   open

   x < 4

4. The statement −2 < x ≤ 1 is read "___ is less than x and x is less than or equal to ___". The graph has an open circle at ___ (because ___ is not a part of the graph) and a solid dot at ___ (because ___ is a part of the graph.

   −2

   1

   −2; −2

   1

   1

   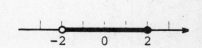

2.9 The Addition and Multiplication Properties of Inequality

Complete each graph.

5. $x < 7$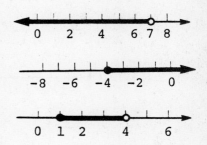

6. $x \geq -4$

7. $1 \leq x < 4$

8. To solve the inequality $x + 12 < 16$, you need to find all values of ____ which make the statement true. To do so, use the addition property of inequality. The inequalities $A < B$ and $A + C <$ _____ have the same _____.

   x

   B + C; solution

9. To solve the inequality $x + 12 < 16$, subtract ____ from both sides.
   $$x + 12 - \_\_\_\_ < 16 - \_\_\_\_$$
   $$x < \_\_\_\_$$

   Draw a graph of this solution on the number line.

   12

   12; 12

   4

   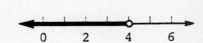

   The open circle on the graph expresses the fact that _____ does not belong to the solution.

   4

10. Solve $x + 8 \geq 5$. Subtract ____ from both sides.
    $$x + 8 - \_\_\_\_ \geq 5 - \_\_\_\_$$
    $$x \geq \_\_\_\_$$

    Graph the solution.

    8

    8; 8

    -3

    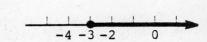

    The heavy dot at _____ shows that -3 (*is/is not*) part of the solution.

    -3; is

11. Solve $3x + 5 \leq 2x - 3$. Subtract 2x from both sides.
    $$3x + 5 - \_\_\_\_ \leq 2x - 3 - \_\_\_\_$$
    $$\_\_\_\_ \leq -3$$

    2x; 2x

    x + 5

Subtract ____ from both sides.

x ≤ ____

Graph the solution.

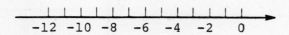

5
−8

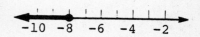

12. Solve $8x + 6 - 3x \leq 3x + 5 + x$.

____ ≤ ____
____ ≤ 5
x ≤ ____

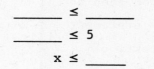

$5x + 6$; $4x + 5$
$x + 6$
−1

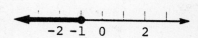

13. Solve $4 + 2x > 3x - 5$.

4 > ____

Add ____ to both sides.

____

$9 > x$ is a correct solution, but it is customary to rewrite it as ____.

$x - 5$
5

$9 > x$

$x < 9$

14. Solve $9 + 4x + 5x - 8 \geq 12x + 4 - 2x$.

____ ≥ ____
1 ≥ ____
____ ≥ x
x ≤ ____

Graph the solution.

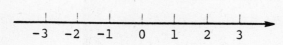

$9x + 1$; $10x + 4$
$x + 4$
−3
−3

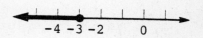

The multiplication property of inequality is similar to the multiplication property of equality, but the direction of the inequality symbol may change.

15. Multiply both sides of the inequality $-6 < 8$ by 4.

( )(−6) < ( )8
−24 ____ 32

Did the direction of the inequality symbol change? (yes/no)

4; 4
<

no

2.9 The Addition and Multiplication Properties of Inequality

16. Multiply both sides of $-6 < 8$ by -3.

    $()(-6) \underline{} 8(-3)$ −3; >

    $18 \underline{} -24$ >

    The direction of the inequality symbol
    (*did/did not*) change.   did

17. By the multiplication property of inequality, if both sides of an inequality are multiplied by a positive number, the direction of the inequality symbol (*does/does not*) change.   does not

    If both sides are multiplied by a negative number, the direction of the inequality symbol (*does/does not*) change.   does

18. Multiply both sides of $6 > -4$ by 9.

    $()6 \underline{} 9(-4)$ 9; >

    $54 \underline{} -36$ >

    The direction of the inequality symbol
    (*did/did not*) change, since 9 is a _____ number.   did not; positive

19. Multiply both sides of $-8 < -2$ by -6.

    $(-6)(-8) \underline{} (-6)(-2)$ >

    $48 \underline{} 12$ >

    The direction here (*did/did not*) change.   did

20. The multiplication property of inequality can be used to solve inequalities. For example, to solve $4m < 12$, divide both sides by ____,   4

    which is a _____ number.   positive

    $\dfrac{4m}{\underline{}} \underline{\phantom{<}} \dfrac{12}{4}$ 4; <

    $m \underline{\phantom{<}} 3$ <

    The direction of the inequality symbol
    (*did/did not*) change.   did not

94 Chapter 2 Solving Equations and Inequalities

21. To solve $3 + 4x - 5x + 6x \leq 23$, first simplify: _____ ≤ 23. Then subtract _____ from both sides: _____ $\leq$ _____. (Adding on both sides (*does/does not*) change the direction of the inequality.) Now divide by _____, which is positive: $x \leq$ _____. Graph the solution.

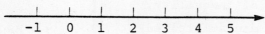

$5x + 3; 3$
$5x; 20$
does not
5
4

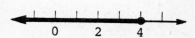

22. To solve $-3x \geq 9$, you would have to divide both sides by _____, which is _____.

$$\frac{-3x}{-3} \quad \underline{\quad} \quad \frac{9}{-3}$$
$$x \leq \underline{\quad}$$

Graph the solution.

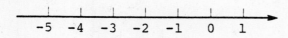

-3; negative
$\leq$
-3

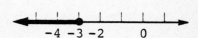

23. Solve $4x - 8x + 6 < 13 + 1$.

_____ < _____
$-4x <$ _____

Divide both sides by _____.

$$\frac{-4x}{-4} \quad \underline{\quad} \quad \frac{8}{-4}$$
$$x > \underline{\quad}$$

Graph the solution.

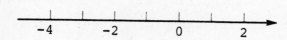

$-4x + 6; 14$
8
-4
$>$
-2

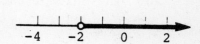

24. Solve $2x - 5x + 6 + 8 - 4x > 2x - 5 + x - 1$.

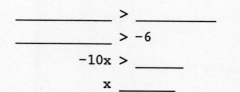

Graph the solution.

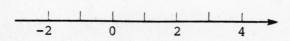

$-7x + 14; 3x - 6$
$-10x + 14$
-20
< 2

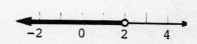

2.9 The Addition and Multiplication Properties of Inequality

25. Solve $4x - 9 - 8x + 5x + 2 \leq 3x - 3$.

    Solution: _____ $x \geq -2$

26. Solve $3r - 8r + 6 - 4 + 2r \geq 3r - 5$.

    Solution: _____ $r \leq 7/6$

27. Inequalities can be used to solve applied problems. For example, "the sum of a number and 4 is less than 11" would be written

    $x +$ ____ $<$ ____ .   4; 11

    Solve this inequality.

    x _____ < 7

Solve each applied problem. Use x as the variable.

28. The difference between five times a number and four times the number is at least 10. Find the possible numbers.

    "At least" is written ____ . $\geq$
    Write the inequality.

    _____ $\geq$ _____ $5x - 4x \geq 10$
    x _____ ≥ 10

29. Two sides of a triangle are 12 inches long and 18 inches long. The perimeter of the triangle must be at least 52 inches. Find the possible lengths for the third side.

    _____   22 inches or longer

30. The perimeter of a triangle must be no more than 21 centimeters. Two of the sides are 8 centimeters in length. Find the possible lengths of the third side.

    _____   5 centimeters or shorter

Chapter 2 Solving Equations and Inequalities

31. Melissa has scores of 87, 84, 95, and 92 on four quizzes. What possible scores can she make on the fifth quiz to have an average of at least 90 after five quizzes?

    Average   is at least   90.
    ↓   ↓   ↓

    x _____ _____

Melissa must score ____ or more.

$\frac{87+84+95+92+x}{5} \geq 90$

≥ ; 92

92

32. Twice a number is at least 50. What are the possible values of the number? _____

25 or more

33. The perimeter of a rectangle cannot exceed 24 meters. The length is 5 meters more than the width. Find the possible values of the width.

    _____

7/2 meters or less

34. If 2/3 of a number is added to 4, the result is less than 2. Find all possible values of the number.

    Any number less than _____

−3

35. The base of a triangle is 7 centimeters. The area must be at least 42 square centimeters. Find the possible values for the height of the triangle.

    (Use the formula _____.)

    _____

$A = \frac{1}{2}bh$

12 centimeters or longer

36. The inequality 5 < x < 9 says that x is _____ 5 and 9.

between

2.9 The Addition and Multiplication Properties of Inequality

Solve each inequality. Graph each solution.

37. Solve $0 < x - 2 < 5$.

Add ____ to each part.   2

$0 +$ ____ $< x - 2 +$ ____ $< 5 +$ ____   2; 2; 2

____ $<$ ____ $<$ ____   2; x; 7

Graph the solution.

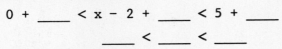
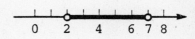

38. Solve $5 \leq x + 7 < 11$.

Solution: ____ $-2 \leq x < 4$

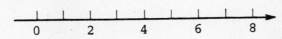

39. $-3 \leq x + 5 < 1$

Solution: ____ $-8 \leq x < -4$

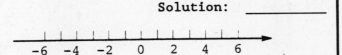

40. The inequality $2 \leq 4x \leq 9$ means that $4x$ is

_____ 2 and 9. To solve this inequality,   between

_____ each part by ____.   divide; 4

$$\frac{2}{4} \leq \frac{4x}{4} \leq \_\_\_$$ $\frac{9}{4}$

or ____ $\leq$ ____ $\leq \frac{9}{4}$. $\frac{1}{2}$; x

Solve each inequality.

41. $3 \leq 5x - 1 \leq 9$

First, add ___ to each part.   1

$3 +$ ____ $\leq 5x - 1 +$ ____ $\leq 9 +$ ____   1; 1; 1

____ $\leq$ ____ ≤ 10 4; 5x

Now divide by ____.   5

____ $\leq x \leq$ ____   4/5; 2

42. $-1 \leq \frac{2}{3}m - 2 \leq 4$ Solution: ____ $3/2 \leq m \leq 9$

43. $-9 < \frac{1}{4}k + 2 < 7$ Solution: ____ $-44 < k < 20$

Chapter 2 Test

The answers for these questions are at the back of this Study Guide.

Simplify by combining like terms.

1. $12q - 5q + 3q - q - q$ 1. _____

2. $2p - 8p + 3 + (-4) - (-2) + 3p$ 2. _____

3. $7(2 - z) - (3z + 4) - 2(5z - 1)$ 3. _____

Solve each equation.

4. $6r + 5 = 2 - 3(r + 2)$ 4. _____

5. $3q - q - 2(q - 4) = -(q + 1)$ 5. _____

6. $\frac{8}{3}p = -24$ 6. _____

7. $\frac{1}{4}r + \frac{1}{2} = \frac{5}{4}r - \frac{1}{3}$ 7. _____

8. $-(z + 2) - 3z + 1 = 2 - 6z - 3$ 8. _____

Write an equation for each of the following, and then solve it. Use x as the variable.

9. Four times a number is subtracted from 5, giving a result of 17. Find the number.   9. _____

10. There were 9 more women than men at a meeting, with 71 people present in total. How many men were present?   10. _____

11. Solve the formula $A = \frac{1}{2}bh$ for b. 11. _____

12. Solve the formula $P = a + b + c$ for b. 12. _____

13. Is the proportion $\frac{17}{35} = \frac{357}{735}$ true or not? 13. _____

Solve each proportion.

14. $\frac{m}{12} = \frac{7}{60}$ 14. _____

15. $\frac{r - 6}{r + 2} = \frac{1}{5}$ 15. _____

16. If 15 tee shirts cost $142.50, find the cost of 7 shirts. 16. _____

Write an equation for each problem, and then solve it.

17. Stacy has some ten-dollar bills and some twenty-dollar bills. She has 7 more twenties than tens. Altogether the money is worth $590. Find the number of ten-dollar bills that she has. 17. _____

18. Two trains leave the same station at the same time, traveling in opposite directions. One train travels at 45 miles per hour, and the other train travels at 65 miles per hour. After how many hours will the trains be 330 miles apart? 18. _____

19. A person invests some money at 12%, with $8000 more than this amount invested at 14%. The total annual income from interest is $5020. How much is invested at each rate? 19. _____

Chapter 2 Solving Equations and Inequalities

20. How many liters of a 40% chemical solution must be mixed with 75 liters of a 70% solution to get a 50% mixture?

20. _____

Solve each inequality.

21. $-5q \geq 30$

21. _____

22. $7(z - 1) + 2 \leq 5(z - 1) + 3z$

22. _____

23. $-2m + 5(m - 1) < 3m - (2 - m)$

23. _____

24. $-1 \leq 5r + 4 < 13$

24. _____

25. Rob Harris scored 84 and 92 on his first two chemistry tests. What must he score on his third test to have an average of at least 90?

25. _____

CHAPTER 3 POLYNOMIALS AND EXPONENTS

3.1 Polynomials

[1] Identify terms and coefficients. (See Frames 1-6 below.)

[2] Add like terms. (Frames 7-9)

[3] Know the vocabulary for polynomials. (Frames 10-17)

[4] Evaluate polynomials (Frames 18-20)

[5] Add polynomials. (Frames 21-31)

[6] Subtract polynomials. (Frames 32-44)

---

1. The expression $8y$, $-9z^4$, and $11k^2$ are examples of _____.	terms
2. The number in front of a term is called a _____.	coefficient
Give the coefficient of each terms.	
3. $8y$ _____	8
4. $-9z^4$ _____	-9
5. $11k^2$ _____	11
6. r _____	1
7. Terms which contain exactly the same _____ raised to exactly the same _____ or with the same _____ are called _____ terms. Write *like terms* or *unlike terms* for each pair of terms.	variables powers exponents; like
$8m^5$ and $-6m^5$ _____	like terms
$-9y^4$ and $-9y^3$ _____	unlike terms
$8x^2$ and $-9x^2$ _____	like terms
$11m^4n^3$ and $-5m^4n^3$ _____	like terms
$6x$ and $6 + x$ _____	unlike terms

Chapter 3 Polynomials and Exponents

8. Only like terms can be combined, that is, added or _____. Like terms are combined by using the _____ property, for example;

    $6m + 5m = ($ ____$)m =$ _____.

    subtracted

    distributive

    $6 + 5$; $11m$

9. Find the sums or differences.

    $3p - 5p = ($ ____$)p =$ _____ $3 - 5$; $-2p$

    $5m - 4m + 3m + 5m = [5 + (-4) +$ ___ $+$ ___$]$ ___ 3; 5; m
    $=$ ____ $9m$

    $8p^3 + 5p^3 =$ _____ $13p^3$

    $-7x^5 + 3x^5 - 5x^5 + 4x^5 =$ _____ $-5x^5$

    $9p + 9q$ (can/cannot) be further combined, since $9p$ and $9q$ are _____ terms.

    cannot

    unlike

10. A finite sum of terms of the form ax^n is called a _____ _____ _____. For example, $8x^4 - 5x^3 + 3x^2$ is a polynomial of ___ terms.

    polynomial in x

    3

    Write the number of terms in each polynomial.

$9m^2 + 4$	2
$13p^5 - 5p^4 + 6p^2 - 5p + 5$	5
$3x^5 + 2x^4 - 5x^6 + 8x^3 - 7x + 1$	6
$8m^3n^4$	1
$9y^3 + 7y^2 - 4y + 8$	4

11. The polynomial $18x^3 - 4x^3 + 6x^2 + 3x - 5$ can be simplified:

    $18x^3 - 4x^3 + 6x^2 + 3x - 5 =$ ____ $+ 6x^2 + 3x - 5$ $14x^3$

    This is a polynomial of ___ terms.   4

12. A polynomial of exactly one term is called a _____. $8x^3$, $-5m^4$, $2x^7$, and $8x^3$ are all examples of _____.

    monomial

    monomials

13. A polynomial of exactly two terms is called a
_____. $-8x^2 + 4x$, $6m^3 + 2m$, $8x + 3$, and
$2y - 5$ are all examples of _____.

binomial

binomials

14. A polynomial of exactly three terms is called a
_____. $-8p^3 + 6p^2 + 5$ is an example of a
_____.

trinomial

trinomial

15. There is no special name for a polynomial of
more than _____ terms.
Write *monomial*, *binomial*, or *trinomial* for each
polynomial. It it is none of these, write
polynomial. First simplify, if possible.

three

$8p^4 - 5p^2 + 6p$ _____

$9m^2 - 4m$ _____

$6 + 8p$ _____

$8r^3 + 6r^3 - 5r^3 + 3r^3$ _____

$5y^5 - 8y^4 + 3y^2 - 5y^2 + 6y^3 + 5$ _____

-9 _____

trinomial

binomial

binomial

monomial
(simplify)

polynomial

monomial

16. In the term (or _____) $8m^5$, the number 8 is
called the _____. Write the coefficient
of the x^2 term for each polynomial.

monomial

coefficient

$8x^2 - 5x + 3$ _____

$9x^3 - 2x^2 - 3x^2 + 5x - 8$ _____

$-9x^5 + 3x^2 - 8x^2 - 9x^2 + 6x - 4$ _____

$12x^4 - 8x^3 + 3x - 7$ _____

8

-5 (Did you
   simplify?)

-14

0

17. The highest _____ in a polynomial of one
variable is called the _____ of the polynomial. Write the degree of each polynomial.

exponent

degree

$8m^4 - 5m^2 - 8m + 3$ _____

$6p^3 - 5p^7 + 8p^2 - 9$ _____

12 _____

$7x + 8$ _____

4

7

0
(Since $12 = 12x^0$)

1

Chapter 3 Polynomials and Exponents

18. If you are given a polynomial, such as $5x^3 + 2x^2 + 3x - 6$, you can find the value of the polynomial for various _____ of x. For example, if $x = -1$, then

    $5x^3 + 2x^2 + 3x - 6$
    $= 5(\quad)^3 + 2(\quad)^2 + 3(\quad) - 6$
    $= 5(\quad) + 2(\quad) - 3 - 6$
    $= \underline{\quad}.$

    values

    $-1; -1; -1$
    $-1; 1$
    -12

    Evaluate this same polynomial for $x = 2$.

    $5x^3 + 2x^2 + 3x - 6$
    $= 5(\quad)^3 + 2(\quad)^2 + 3(\quad) - 6$
    $= 5(\quad) + 2(\quad) + 6 - 6$
    $= \underline{\quad}$

    $2; 2; 2$
    $8; 4$
    48

19. Evaluate $-4x^2 + 11x - 6$ for $x = -3$.

    $-4x^2 + 11x - 6 = -4(\quad)^2 + 11(\quad) - 6$
    $= \underline{\quad}$

    $-3; -3$
    -75

20. Evaluate $3x^2 + 8x + 9$ for $x = -6$. _____

    69

21. Warm-up exercises: Add the terms.

    $3p^2 + 8p^2 = (\quad)p^2 = \underline{\quad}$
    $-5p + 7p = (\quad)p = \underline{\quad}$
    $9r^3 + 2r^3 = (\quad)r^3 = \underline{\quad}$
    $-4r^2 + 7r^2 = \underline{\quad}$ Terms are (*like/unlike*).
    $2m^5 + 3m^5 = \underline{\quad}$ Terms are (*like/unlike*).
    $-4m^3 + 6m^2 = \underline{\quad}$ Terms are (*like/unlike*).

    $3 + 8; 11p^2$
    $-5 + 7; 2p$
    $9 + 2; 11r^3$
    $3r^2$; like
    $5m^5$; like
    $-4m^3 + 6m^2$; unlike

22. Add the polynomial $3p^2 - 5p$ and the polynomial $7p - 5$.

    $(3p^2 - 5p) + (7p - 5) = 3p^2 + (-5p + \underline{\quad}) - 5$
    $= 3p^2 + (\quad)p - 5$
    $= 3p^2 + \underline{\quad} - 5$

    $7p$
    $-5 + 7$
    $2p$

3.1 Polynomials

Add: $(3p^2 - 5p + 2) + (8p^2 + 7p - 5)$
= $(3p^2 + \_\_\_) + (-5p + \_\_\_) + (2 - 5)$ $8p^2$; $7p$
= ___ + ___ − ___. $11p^2 + 2p - 3$

To add two polynomials, add the ____ terms in the polynomials.   like

23. $(9r^3 - 4r^2 + 6r) + (2r^3 + 7r^2 - 9r)$
= $(9r^3 + \_\_\_) + (-4r^2 + \_\_\_) + (6r \_\_\_)$ $2r^3$; $7r^2$; $-9r$
= ___ + ___ − ___ $11r^3$; $3r^2$; $3r$

24. $(2m^5 - 4m^3 - 8m^2) + (3m^5 - 9m^3 + 6m^2)$
= ___ − ___ − ___ $5m^5$; $13m^3$; $2m^2$

25. $(4x^3 - 8x^2 + 5) + (3x^3 + 11x^2 - 9)$
= _____ $7x^3 + 3x^2 - 4$

26. $(8p^4 - 9p^2 + 6) + (3p^3 + 11p^2 - 8)$
= _____ $8p^4 + 3p^3 + 2p^2 - 2$

27. Polynomials can also be added by placing one _____ the other, with _____ terms lined up in _____. Add the polynomials.   above; like columns

$$6m^4 - 8m^2 + 3m + 5$$
$$-2m^4 + 6m^2 - 5m + 8$$

The sum of the polynomials is _____ $4m^4 - 2m^2 - 2m + 13$

28.
$$6p^3 - 4p^2 + 8p - 5$$
$$-7p^3 + 5p^2 + 9p + 11$$

Sum: $-p^3 + p^2 + 17p + 6$

29.
$$-3m^4 + 6m^3 - 9m^2 + 8m - 5$$
$$7m^4 - 9m^3 + 10m^2 - 8m + 12$$

Sum: $4m^4 - 3m^3 + m^2 + 7$

Chapter 3 Polynomials and Exponents

30. $12y^5 + 8y^4 + 6y^2$
   $\underline{-15y^5 - 9y^4 + 12y^3 - 8y^2 + 4}$

   Sum: _____ $-3y^5 - y^4 + 12y^3 - 2y^2 + 4$

31. $8r^6 - 4r^3 + 6$
   $\underline{ 9r^5 + 4r^2 - 9}$

   Sum: _____ $8r^6 + 9r^5 - 4r^3 + 4r^2 - 3$

32. To subtract two polynomials, _____ the signs   change
    of the second polynomial and ____. To work the   add
    problem $(8x^2 - 4x - 5) - (2x^2 + 3x - 7)$, change
    the _____ on the _____ polynomial, and   signs; second
    _____ like terms.   add

    $(8x^2 - 4x - 5)$ ___ $(-2x^2 - $ _____$)$ +; $3x + 7$

    $=$ _____ $6x^2 - 7x + 2$

33. $(5a - 7b) - (-3a + 4b)$

    $= (5a - 7b)$ ___ (_____)   +; $3a - 4b$

    $=$ _____ $8a - 11b$

34. $(9r^3 - 8r^2 - 7) - (-2r^3 + 3r^2 + 5)$

    $= (9r^3 - 8r^2 - 7)$ ___ (_____)   +; $2r^3 - 3r^2 - 5$

    $=$ _____ $11r^3 - 11r^2 - 12$

35. $(2y^3 - 8y^2 - 9y + 2) - (-8y^3 + 6y^2 - 4y)$

    $=$ _____ $10y^3 - 14y^2 - 5y + 2$

36. $(9p^2 - 11p - 8) - (-4p^2 - 5p - 8)$

    $=$ _____ $13p^2 - 6p$

37. $(-12x^3 - 4x^2 + 6) - (3x^3 - 11x^2 + 5x)$

    $=$ _____ $-15x^3 + 7x^2 - 5x + 6$

38. Subtraction can also be worked in columns. Here we change the signs of the (*top/bottom*) polynomial, and _____.

bottom

add

$$9m^4 - 8m^2 + 6m - 3$$
$$-4m^4 + 6m^2 + 11m - 8$$

Difference:

$13m^4 - 14m^2 - 5m + 5$

39.
$$-12p^3 - 8p^2 + 7p + 2$$
$$3p^3 + 9p^2 - 6p + 4$$

Difference:

$-15p^3 - 17p^2 + 13p - 2$

40.
$$8r^2 - 11r + 6$$
$$-4r^2 + 14r + 6$$

Difference:

$12r^2 - 25r$

41.
$$18p^6 - 4p^3 + 11p^2 - 6p + 8$$
$$12p^6 - 4p^3 + 9p^2 - 4p + 8$$

Difference:

$6p^6 + 2p^2 - 2p$

Work these problems using both addition and subtraction.

42. $(9m^2 - 5m) + (8m^2 - 6m) - (3m^2 - 2)$

= $9m^2 - 5m + 8m^2 - 6m$ _____

= _____

$- 3m^2 + 2$

$14m^2 - 11m + 2$

43. $[(12x^4 - 9x^2 + 6) - (3x^4 - 8x^2 - 5)] + (x^2 - 4)$

= _____

$9x^4 + 7$

44. $[(y^8 - y^6 + y^2) - (3y^8 + 2y^2)] + (7y^6 + 8y^2)$

= _____

$-2y^8 + 6y^6 + 7y^2$

Chapter 3 Polynomials and Exponents

3.2 Exponents

[1] Use exponents. (See Frames 1-6 below.)

[2] Use the product rule for exponents. (Frames 7-10)

[3] Use the rule $(a^m)^n = a^{mn}$. (Frames 11-15)

[4] Use the rule $(ab)^m = a^m b^m$. (Frames 16-21)

[5] Use the rule $\left(\dfrac{a}{b}\right)^m = \dfrac{a^m}{b^m}$. (Frames 22-26)

---

1. To square a number, multiply the number by itself.

1 squared: $1^2 = 1 \cdot 1 = 1$	
2 squared: $2^2 = 2 \cdot 2 =$ ____	4
3 squared: $3^2 = 3 \cdot$ ____ $= 9$	3
4 squared: $4^2 =$ ____ $\cdot 4 = 16$	4
5 squared: 5——$= 5 \cdot 5 =$ ____	$5^2 = 25$
6 squared: ____ $^2 = 6 \cdot 6 =$ ____	$6^2 = 36$
7 squared: $7^2 =$ ____ $\cdot$ ____ $= 49$	$7 \cdot 7$
8 squared: $8^2 =$ ____ $\cdot 8 = 64$	8
9 ____ : 9——$= 9 \cdot 9 =$ ____	squared; $9^2 = 9 \cdot 9 = 81$
10 squared: _____	$10^2 = 10 \cdot 10 = 100$

   The square of x is $x \cdot x = x$——. x^2

2. To cube a number, multiply it by itself, then multiply by itself again:

    number · number · number.

    Then 1 cubed is $1^3 = 1 \cdot 1 \cdot 1 = 1$.

2 cubed: $2^3 = 2 \cdot 2 \cdot$ ____ $= 8$	2
3 cubed: $3^3 = 3 \cdot$ ____ $\cdot$ ____ $=$ ____	$\cdot 3 \cdot 3$; 27
4 cubed: 4—— $= 4 \cdot 4 \cdot 4 =$ ____	4^3; 64
5 cubed: ____ $^3 =$ ____ $\cdot$ ____ $\cdot$ ____ $= 125$	5^3; $5 \cdot 5 \cdot 5$
6 cubed: ____ $= 6 \cdot 6 \cdot 6 =$ ____	6^3; 216
7 cubed: ____ $=$ ____ $\cdot$ ____ $\cdot$ ____ $= 343$	$7^3 = 7 \cdot 7 \cdot 7$
8 cubed: _____	$8^3 = 8 \cdot 8 \cdot 8 = 512$
9 cubed: _____	$9^3 = 9 \cdot 9 \cdot 9 = 729$

3.2 Exponents

    10 _____: $10^3 = 10 \cdot 10 \cdot 10 = 1000$ | cubed

    x cubed: $x^3 = x \cdot$ _____ $\cdot$ _____ | $\cdot x \cdot x$

3. Then, 5 squared is $5 \cdot 5$, and the shorthand symbol is 5^2. The 2 is an exponent or power. The number that gets multiplied by itself, like 5, is the _____. Together, 5^2 is an exponential expression. You can say "5 squared" or "5 to the second power." You can say "5 cubed" or "5 to the _____ power." Then 5 to the fourth power is $5^4 = 5 \cdot 5 \cdot 5 \cdot 5$; and 5 to the fifth power is

    $5^{—}$ = _____ $\cdot$ _____ $\cdot$ _____ $\cdot$ _____ $\cdot$ _____ .

    | base
    | third
    | 5^5; $5 \cdot 5 \cdot 5 \cdot 5 \cdot 5$

Write 5 to the sixth power: _____. Write 5 to the seventh power: _____.

    | 5^6
    | 5^7

4. Write the base and exponent.

Exponential Expression	Base	Exponent	
1^3	1	_____	3
2^6	_____	6	2
3^5	_____	5	3
7^4	_____	_____	7; 4
k^7	_____	_____	k; 7

5. Write an exponential expression for the given base and exponent. Then write out the operation, and find the value by multiplying.

Base	Exponent	Exponential Expression	Operation	Value	
2	2	2^2	$= 2 \cdot 2$	$= 4$	
4	3	_____	$=$ _____ $\cdot$ _____ $\cdot$ _____	$=$ _____	4^3; $4 \cdot 4 \cdot 4$; 64
10	2	_____	$=$ _____	$=$ _____	10^2; $10 \cdot 10$; 100
3	4	_____	$=$ _____	$=$ _____	3^4; $3 \cdot 3 \cdot 3 \cdot 3$; 81
6	2	_____	$=$ _____	$=$ _____	6^2; $6 \cdot 6$; 36
1	5	_____	$=$ _____	$=$ _____	1^5; $1 \cdot 1 \cdot 1 \cdot 1 \cdot 1$; 1

If $x = 3$, what is the value of x^2? _____ If $k = 5$, what is the value of k cubed? _____

    | 9
    | 125

Chapter 3 Polynomials and Exponents

6. Write a shorthand form of the expression on the left.

	With Exponents	Read as:	
$7 \cdot 7$	$7^{\_\_}$	$\_\_$ squared	7^2; seven
$2 \cdot 2 \cdot 2 \cdot 2$	$\_\_$	2 to the $\_\_$	2^4; fourth
$m \cdot m \cdot m$	$\_\_$	m $\_\_$	m^3; cubed
The area of a a square whose side measures 8	$\_\_$	8 squared is $\_\_$.	8^2; 64

7. To multiply exponential expressions, make sure that the expressions have the same base. In the product $2^3 \cdot 2^4$, the expressions have (*the same/ not the same*) bases. Write *same* or *not same* for the bases in each product.

the same

$9^2 \cdot 9^4$	$\_\_$	$12^5 \cdot 2^{12}$	$\_\_$	same; not same
$2^5 \cdot 2^{12}$	$\_\_$	$7 \cdot 7^4$	$\_\_$	same; same
$4^8 \cdot 4^5$	$\_\_$	$6^6 \cdot 3^3$	$\_\_$	same; not same
$8^{12} \cdot 8^{17}$	$\_\_$	$x^2 \cdot x^3$	$\_\_$	same; same
$9^2 \cdot 8^2$	$\_\_$	$m^{12} \cdot m^{15}$	$\_\_$	not same; same

Which of the above *cannot* be multiplied using the product rule for exponents?

$_____$

$12^5 \cdot 2^{12}$; $6^6 \cdot 3^3$; $9^2 \cdot 8^2$ (Bases are not same)

8. Find the product of $2^4 \cdot 2^7$. The bases are (*the same/not the same*). Add the exponents: $4 + 7 = \_\_$.

the same

11

Multiply: $2^4 \cdot 2^7 = 2^{\_+\_} = 2^{11}$

$4 + 7$

Multiply: $3^3 \cdot 3^2 = 3^{3+2} = 3^{\_}$.

3^5

3.2 Exponents

9. To get the product of exponential expressions, _____ exponents if the bases (*are/are not*) the same. This rule is called the _____ rule. You multiply 3^4 by 3^5 to get 3^9 by the _____ rule. Can you multiply 3^2 by 4^2 using the product rule? (*yes/no*). Simplify the following by the product rule.

	add; are
	product
	product
	no

$2^3 \cdot 2^8 = 2^{\text{---}+8} = 2^{\text{---}}$ 2^{3+8}; 2^{11}

$2^5 \cdot 2^{12} = $ _____ 2^{17}

$3^4 \cdot 3^5 = 3^{4+\text{---}} = 3^{\text{---}}$ 3^{4+5}; 3^9

$5^2 \cdot 5^2 = $ _____ 5^4

$11^3 \cdot 11^8 = 11^{\text{---}}$ 11^{11}

$2^3 \cdot 2^4 \cdot 2^2 = 2^{\text{---}+\text{---}+\text{---}} = $ _____ 2^{3+4+2}; 2^9

$9^4 \cdot $ _____ $^9 = 9^{13}$ 9^9

$3^4 \cdot 3^4 \cdot 3 = 3^{4+4+\text{---}} = $ _____ 3^{4+4+1}; 3^9

$7^3 \cdot 7^0 = 7^{\text{---}}$ 7^3

$(6^2 \cdot 6^3)(6^7) = $ _____ 6^{12}

10. In summary, the product rule for exponents says
    $a^m \cdot a^n = $ _____. a^{m+n}

11. By power rule (a) for exponents,
    $(a^m)^n = $ _____. a^{mn}

12. The expression 3^2 to the fourth power is written with parentheses as $(3^2)^4$. Multiply exponents.

    $(3^2)^4 = 3^{2 \cdot \text{---}} = 3^8$ $3^{2 \cdot 4}$; 3^8

    $(3^2)^5 = 3^{\text{---} \cdot \text{---}} = 3^{\text{---}}$ $3^{2 \cdot 5}$; 3^{10}

Use power rule (a) to simplify.

13. $(2^5)^3 = $ _____ 2^{15}

14. $(3^8)^2 = $ _____ 3^{16}

15. $(4^6)^5 = $ _____ 4^{30}

112   Chapter 3   Polynomials and Exponents

16. By power rule (b),

$(ab)^m = $ _____. | $a^m b^m$

Use power rule (b) to simplify.

17. $(5pq)^3 = $ _____ | $5^3 p^3 q^3$ or $125 p^3 q^3$

18. $6(zy)^4 = $ _____ | $6z^4 y^4$

19. $(3r^4 z^2)^3 = $ _____ | $3^3 r^{12} z^6$ or $27 r^{12} z^6$

20. $(5y^2)^3 = $ _____ | $125 y^6$

21. $(9z^4 x^7)^2 = $ _____ | $81 z^8 z^{14}$

22. We have a similar rule for quotients. It is called power rule (c).

$\left(\dfrac{a}{b}\right)^m = $ _____.   (if $b \neq 0$). | $\dfrac{a^m}{b^m}$

Use power rule (c) to simplify.

23. $\left(\dfrac{3}{k}\right)^7 = $ _____ | $\dfrac{3^7}{k^7}$

24. $\left(\dfrac{8^3}{z^2}\right)^4 = $ _____ | $\dfrac{8^{12}}{z^8}$

25. $\left(\dfrac{-3}{8}\right)^2 = $ _____ | $\dfrac{9}{64}$

26. $-\left(\dfrac{2x}{3y^2}\right)^3 = $ _____ | $-\dfrac{8x^3}{27 y^6}$

3.3 Multiplication of Polynomials

[1] Multiply a monomial and a polynomial. (See Frames 1–6 below.)

[2] Multiply two polynomials. (Frames 7–25)

[3] Multiply binomials by the FOIL method. (Frames 26–39)

3.3 Multiplication of Polynomials   113

1. Warm-up exercises: multiply.

$4 \cdot x =$ ___	$x \cdot x =$ ___	$4x$; x^2
$7 \cdot m =$ ___	$6x \cdot x =$ ___	$7m$; $6x^2$
$6 \cdot y^2 =$ ___	$k \cdot 4k =$ ___	$6y^2$; $4k^2$
$x \cdot 2 =$ ___	$p^2 \cdot 5 =$ ___	$2x$; $5p^2$
$2x \cdot 3x = (2 \cdot 3)(x^{1+}\text{---}) =$ ___		$(2\cdot 3)(x^{1+1})$; $6x^2$
$5y^2 \cdot 2y^2 = ($___$\ 2)(y^{2+}\text{---}) =$ ___		$(5\cdot 2)(y^{2+2})$; $10y^4$
$2m^2 \cdot 3m^3 =$ ___		$6m^{2+3} = 6m^5$
$4(x + 1) = 4x +$ ___		4
$x(x + 2) = x^2 +$ ___		$2x$
$2m(m - 4) =$ ___		$2m^2 - 8m$
$3p(4p^2 - p) =$ ___		$12p^3 - 3p^2$
$5p^2(2p^3 - 2p) =$ ___		$10p^5 - 10p^3$

2. Multiply:

$$2(300 + 40 + 6) = 2(300) + 2(\ ) + \underline{\ \ }(6)$$ 40; 2
$$= \underline{\ \ } + \underline{\ \ } + \underline{\ \ }$$ 600; 80; 12
$$= \underline{\ \ }$$ 692

Multiply:

$$3x(4x^2 - 11x + 4) = 3x(\ ) + (\ )(-11x) + \underline{\ \ }$$ $4x^2$; $3x$; $12x$
$$= \underline{\ \ \ \ \ \ } - \underline{\ \ \ \ \ \ } + 12x$$ $12x^3$; $33x^2$

To multiply a polynomial by a monomial (a monomial is a polynomial of ___ term), multiply   one
each ___ of the polynomial by the monomial.   term

3. $9m(2m^3 - 4m^2 + 6m)$
   $= (\ )(2m^3) + (9m)(\ ) + (9m)(\ )$ $9m$; $-4m^2$; $6m$
   $=$ _____ $18m^4 - 36m^3 + 54m^2$

4. $2p(8p^3 - 9p^2 + 6p) =$ _____ $16p^4 - 18p^3 + 12p^2$

5. $-4r^2(2r^4 - 8r^2 + 7r)$
   $= (-4r^2)(\ ) + (-4r^2)(\ ) + (-4r^2)(\ )$ $2r^4$; $-8r^2$; $7r$
   $=$ _____ $-8r^6 + 32r^4 - 28r^3$

114   Chapter 3   Polynomials and Exponents

6. $-9x^2(3x^4 - 5x^2 + 11x - 2)$
   $= -27x^6$ _____ | $+ 45x^4 - 99x^3 + 18x^2$

7. The _____ property is used repeatedly to find the product of any two polynomials. | distributive

8. To find the product of the polynomials $3x + 1$ and $2x - 4$ use the distributive property in the following way.

   $(3x + 1)(2x - 4)$
   $= (\ \ )(2x - 4) + (\ \ )(2x - 4)$ | $3x;\ 1$
   $= (3x)(2x) + 3x(\ \ ) + (1)(\ \ ) + (1)(\ \ )$ | $-4;\ 2x;\ -4$
   $= (\ \ ) - 12x + 2x - (\ \ )$ | $6x^2;\ 4$
   $= 6x^2 + \underline{\ \ \ } - 4$ | $-10x$

Multiply the following polynomials.

9. $(2x + 1)(4x - 3) = (\ \ )(4x - 3) + (\ \ )(4x - 3)$ | $2x;\ 1$
   $= \underline{\ \ \ \ \ \ \ \ }$ | $8x^2 - 2x - 3$

10. $(6x - 2)(x + 1) = \underline{\ \ \ \ \ \ \ \ }$ | $6x^2 + 4x - 2$

11. $(x + 1)(x - 1) = \underline{\ \ \ \ \ \ \ \ }$ | $x^2 - 1$

12. Find the product of $y^2 - 1$ and $y^2 + y + 1$.
    $(y^2 - 1)(y^2 + y + 1)$
    $= (\ \ )(y^2 + y + 1) - (\ \ )(y^2 + y + 1)$ | $y^2;\ 1$
    $= y^4 + \underline{\ \ \ } + y^2 - y^2 - \underline{\ \ \ } - 1$ | $y^3;\ y$
    $= \underline{\ \ \ \ \ \ \ \ }$ | $y^4 + y^3 - y - 1$

Multiply the following polynomials.

13. $(6x + 1)(x^2 + 2x - 1) = \underline{\ \ \ \ \ \ \ \ }$ | $6x^3 + 13x^2 - 4x - 1$

14. $(2a - 1)(a^3 + 2a^2 + a - 6) = \underline{\ \ \ \ \ \ \ \ }$ | $2a^4 + 3a^3 - 13a + 6$

15. $(x^2 + 2x + 1)(x^2 - 2x + 1) = \underline{\ \ \ \ \ \ \ \ }$ | $x^4 - 2x^2 + 1$

3.3 Multiplication of Polynomials 115

16. To find the product $(2r - 3)(3r + 1)(r - 4)$, first multiply _____ and _____. $2r - 3$; $3r + 1$

    $(2r - 3)(3r + 1) = $ _____. $6r^2 - 7r - 3$

    Now multiply:

    $(2r - 3)(3r + 1)(r - 4)$
    $= ($ _____ $)(r - 4)$ $(6r^2-7r-3)(r-4)$
    $= $ _____ $6r^3-31r^2+25r+12$

17. Multiply: $(5p - 2)(3p + 2)(p - 2)$

    _____ $15p^3-26p^2-12p+8$

18. To multiply $3x - 2$ and $x^2 - 4x + 5$, proceed as follows.

    Step 1: Write the problem vertically.

    $x^2 - 4x + 5$
    _____ $3x - 2$

    Step 2: Multiply ____ times each term in the top row. -2

    $x^2 - 4x + 5$
    $ 3x - 2$

    _____ $-2x^2 + 8x - 10$

    Step 3: Multiply ____ times the top row. $3x$

    $x^2 - 4x + 5$
    $ 3x - 2$
    $\overline{}$
    $-2x^2 + 8x - 10$
    _____ $3x^3 - 12x^2 + 15x$

    Step 4: ____ like terms. Add

    $x^2 - 4x + 5$
    $ 3x - 2$
    $\overline{}$
    $ -2x^2 + 8x - 10$
    $3x^3 - 12x^2 + 15x$

    _____ $3x^3-14x^2+23x-10$

Chapter 3 Polynomials and Exponents

Multiply the polynomials in Frames 19–25.

19. $-2y^2 + 3y - 7$
     $ 2y - 5$
     _____

$10y^2-15y+35$
$-4y^3+6y^2-14y$
$-4y^3+16y^2-29y+35$

20. $6p^3 - 4p^2 + 2p$
     $ 5p - 3$
     _____

$-18p^3+12p^2-6p$
$30p^4-20p^3+10p^2$
$30p^4-38p^3+22p^2-6p$

21. $2m^3 - 3m^2 + 4$
     $2m - 3$
     _____

$-6m^3+9m^2-12$
$4m^4-6m^3+8m$
$4m^4-12m^3+9m^2+8m-12$

22. $6k - 1$
     $\underline{2k + 3}$

     _____ $12k^2 + 16k - 3$

23. $8p - 3$
     $\underline{2p + 5}$

     _____ $16p^2 + 34p - 15$

24. $9m^2 - 4m + 3$
     $\underline{4m - 1}$

     _____ $36m^3 - 25m^2 + 16m - 3$

25. $8r^2 - 3r + 2$
     $\underline{-4r + 3}$

     _____ $-32r^3 + 36r^2 - 17r + 6$

26. To multiply two binomials, use the FOIL method.
    These letters are an abbreviation for _____.   first
    _____, _____, _____ terms.   outer; inner; last

3.3 Multiplication of Polynomials

27. Use FOIL to multiply $p + 2$ and $p + 5$.

F:	$(p + 2)(p + 5)$	$p \cdot p =$ ____	p^2
O:	$(p + 2)(p + 5)$	$p \cdot 5 =$ ____	$5p$
I:	$(p + 2)(p + 5)$	$2 \cdot p =$ ____	$2p$
L:	$(p + 2)(p + 5)$	$2 \cdot 5 =$ ____	10

Product: ___ + ___ + ___ + ___ p^2; $5p$; $2p$; 10

= _____ $p^2 + 7p + 10$

Find each of the following products.

28. $(2x - 3)(x + 6)$

F: _____ = ____ $2x \cdot x$; $2x^2$

O: _____ = ____ $2x \cdot 6$; $12x$

I: _____ = ____ $-3 \cdot x$; $-3x$

L: _____ = ____ $-3 \cdot 6$; -18

Product: _____ $2x^2 + 9x - 18$

29. $(5m + 3)(4m + 1)$

F: ___ O: ___ I: ___ L: ___ $20m^2$; $5m$; $12m$; 3

Product: _____ $20m^2 + 17m + 3$

30. $(4k + 5)(5k - 7)$

F: ___ O: ___ I: ___ L: ___ $20k^2$; $-28k$; $25k$; -35

Product: _____ $20k^2 - 3k - 35$

31. $(5r - 9s)(2r + s)$

F: ___ O: ___ I: ___ L: ___ $10r^2$; $5rs$; $-18rs$; $-9s^2$

Product: _____ $10r^2 - 13rs - 9s^2$

Find each product, using FOIL.

32. $(y + 5)(2y - 1) =$ _____ $2y^2 + 9y - 5$

118   Chapter 3   Polynomials and Exponents

33. $(2m + 5)(3m + 7) = $ _____ $6m^2 + 29m + 35$

34. $(3r - 4)(2r - 5) = $ _____ $6r^2 - 23r + 20$

35. $(2x + 4)(3x + 5) = $ _____ $6x^2 + 22x + 20$

36. $(4k - 3)(2k - 8) = $ _____ $8k^2 - 38k + 24$

37. $(5b + 2)(6b - 9) = $ _____ $30b^2 - 33b - 18$

38. $(7y - z)(2y + 3z) = $ _____ $14y^2 + 19yz - 3z^2$

39. $(2k + 5m)(6k - 7m) = $ _____ $12k^2 + 16km - 35m^2$

3.4   Special Products

[1] Square binomials. (See Frames 1–6 below.)

[2] Find the product of the sum and difference of two terms. (Frames 7–12)

[3] Find higher powers of binomials. (Frames 13–18)

1. To square a binomial, you have to multiply.

   $(4r - 5)^2 = ($ $)($ $)$ $4r - 5;\ 4r - 5$

   $ = $ ___ + ___ + ___ + ___ $16r^2;\ (-20r);$
   $(-20r);\ 25$

   $ = $ _____ $16r^2 - 40r + 25$

2. $(2a + 3)^2 = ($ $)($ $)$ $2a + 3;\ 2a + 3$

   $ = $ ___ + 2( ) + ___ $4a^2;\ 6a;\ 9$

   $ = $ _____ $4a^2 + 12a + 9$

3. $(6p - 4)^2 = $ _____ $36p^2 - 48p + 16$

4. $(3x + 2)^2 = $ _____ $9x^2 + 12x + 4$

5. $(4k - 3m)^2 = $ _____ $16k^2 - 24km + 9m^2$

3.4 Special Products   119

6. $(5r + 6s)^2 =$ _____ | $25r^2 + 60rs + 36s^2$

7. A product of the sum and difference of two terms, of the form $(x + y)(x - y)$, is the difference of two squares, _____. The product, $(2a + 3)(2a - 3)$ is _____. Also, $(x + 2)(x$ ___ $2) = x^2 - 4$ and $(y - 3)(y +$ ___$) = y^2 - 9$. | $x^2 - y^2$; $4a^2 - 9$; $-$; 3

8. $(2m + 5)(2m - 5) =$ _____ | $4m^2 - 25$

9. $(3a - 1)(3a + 1) = 9a^2$ _____ | $- 1$

10. $(7r + 3)($ $) = 49r^2 - 9$ | $7r - 3$

11. $(11y - 5)(11y + 5) =$ _____ | $121y^2 - 25$

12. Find the products.

$(6y - 11)(y + 2) =$ _____ | $6y^2 + y - 22$
$(6p - 5)^2 =$ _____ | $36p^2 - 60p + 25$
$(3r - 4)(3r + 5) =$ _____ | $9r^2 + 3r - 20$
$(4k - 3)(4k + 3) =$ _____ | $16k^2 - 9$
$(5x - 4)(x + 5) =$ _____ | $5x^2 + 21x - 20$
$(5x + 4)(5x + 4) =$ _____ | $25x^2 + 40x + 16$

Use the methods in the previous section and this section to find higher powers of binomials.

13. $(a + 3)^3 = (a + 3)^2$ _____
    $=$ _____ $(a + 3)$
    $=$ _____ | $(a + 3)$; $(a^2 + 6a + 9)$; $a^3 + 9a^2 + 27a + 27$

14. $(y - 2)^3 =$ _____ $(y - 2)$
    $=$ _____ $(y - 2)$
    $=$ _____ | $(y - 2)^2$; $y^2 - 4y + 4$; $y^3 - 6y^2 + 12y - 8$

15. $(3r + 2)^3 =$ _____ | $27r^3 + 54r^2 + 36r + 8$

120   Chapter 3   Polynomials and Exponents

16. $(k + 2)^4 = (k^2 + 4k + 4)(\underline{\hspace{2cm}})$ $k^2 + 4k + 4$

         $= 4k^2 + \underline{\hspace{1cm}} + 16$ $16k$

               $+ 4k^3 + \underline{\hspace{1cm}} + 16k$ $16k^2$

               $+ \underline{\hspace{1cm}} + 4k^3 + \underline{\hspace{1cm}}$ $k^4; \; 4k^2$

         $= \underline{\hspace{4cm}}$ $k^4+8k^3+24k^2+32k+16$

17. $(w - 3)^4 = \underline{\hspace{4cm}}$ $w^4 - 12w^3 + 54w^2 - 108w + 81$

18. $(3t - 2)^4 = \underline{\hspace{4cm}}$ $81t^4 - 216t^3 + 216t^2 - 96t + 16$

3.5 The Quotient Rule and Integer Exponents

[1] Use zero as an exponent. (See Frames 1–4 below.)

[2] Use negative numbers as exponents. (Frames 5–7)

[3] Use the quotient rule for exponents. (Frames 8–14)

[4] Use combinations of rules. (Frames 15–26)

[5] Use variables as exponents. (Frames 27–30)

---

1. By definition, a number with exponent 0 equals ____. That is, $a^0 = $ ____, if $a \neq 0$.   1; 1

Evaluate each expression.

2. $6^0 = $ ____   1

3. $(-9)^0 = $ ____   1

4. $-(-15)^0 = $ ____ -1

5. So far, we have discussed only zero exponents and _____ integer exponents. We can also work with negative integer exponents if we make the following definition: positive

3.5 The Quotient Rule and Integer Exponents

$a^{-n} = $ ____.

For example, $3^{-2} = \dfrac{1}{3^{__}} = $ ____.

$\dfrac{1}{a^n}$

$\dfrac{1}{3^2} = \dfrac{1}{9}$

6. Simplify each expression.

$2^{-4} = \dfrac{1}{2^{__}} = $ ____ $\dfrac{1}{2^4} = \dfrac{1}{16}$

$4^{-2} = \dfrac{1}{4^{__}} = $ ____ $\dfrac{1}{4^2} = \dfrac{1}{16}$

$3^{-4} = \dfrac{1}{__} = $ ____ $\dfrac{1}{3^4} = \dfrac{1}{81}$

$5^{-2} = \dfrac{1}{__} = $ ____ $\dfrac{1}{5^2} = \dfrac{1}{25}$

$8^{-1} = \dfrac{1}{__} = $ ____ $\dfrac{1}{8^1} = \dfrac{1}{8}$

$9^{-2} = $ ____ $\dfrac{1}{81}$

$7^{-2} = $ ____ $\dfrac{1}{49}$

$\left(\dfrac{2}{3}\right)^{-2} = \dfrac{1}{\left(\dfrac{2}{3}\right)^{__}} = \dfrac{1}{\dfrac{__}{9}} = $ ____ $\dfrac{1}{\left(\dfrac{2}{3}\right)^2} = \dfrac{1}{\dfrac{4}{9}} = \dfrac{9}{4}$

$\left(\dfrac{3}{4}\right)^{-3} = \dfrac{1}{\left(\dfrac{3}{4}\right)^{__}} = \dfrac{1}{__} = $ ____ $\dfrac{1}{\left(\dfrac{3}{4}\right)^3} = \dfrac{1}{\dfrac{27}{64}} = \dfrac{64}{27}$

$\left(\dfrac{5}{2}\right)^{-2} = $ ____; $\left(\dfrac{2}{9}\right)^{-1} = $ ____; $\left(\dfrac{1}{8}\right)^{-2} = $ ____ $\dfrac{4}{25}$; $\dfrac{9}{2}$; 64

$\left(\dfrac{3}{8}\right)^{-1} = \dfrac{1}{\left(\dfrac{3}{8}\right)^{\_\_}}$ 1

$= \dfrac{1}{\_\_}$ $\dfrac{3}{8}$

$= $ ____ $\dfrac{8}{3}$

7. Expressions with negative exponents can be simplified in much the same way that we simplified expressions with _____ exponents. For example, use the product rule:

    positive

$$2^{-5} \cdot 2^9 = 2^{\_\_} = 2^{\_\_}.$$

    2^{-5+9}; 2^4

Also, $3^{-8} \cdot 3^{-2} \cdot 3^5 = 3^{\_\_} = 3^{\_\_}.$

    3^{-8-2+5}; 3^{-5}

In general, if m and n are any integers,

$$a^m \cdot a^n = a^{\_\_}.$$

    a^{m+n}

Chapter 3 Polynomials and Exponents

8. The quotient rule for exponents says

$$\frac{a^m}{a^n} = \underline{}.$$

a^{m-n}

Evaluate each expression. Write with only positive exponents.

9. $\dfrac{6^{11}}{6^7} = 6^{\underline{}} = \underline{}$

6^{11-7}; 6^4

10. $\dfrac{4^5}{4^3} = \underline{}$

4^{5-3} or 4^2

11. $\dfrac{12^{15}}{12^8} = \underline{}$

12^7

12. $\dfrac{6^4}{6^{-3}} = \underline{}$

$6^{4-(-3)} = 6^7$

[Be careful: $4 - (-3) = 4 + 3 = \underline{}$.]

7

Simplify each expression.

13. $\dfrac{2^{-4}}{2^{-5}} = \underline{}$

$2^{-4-(-5)} = 2^1 = 2$

14. $\dfrac{8^{-4}}{8^{-9}} = \underline{}$

$8^{-4-(-9)} = 8^5$

15. How do you go about simplifying $\dfrac{(3^5)^2 \cdot 3^6}{(3^7)^2}$?

First simplify $(3^5)^2$ as $\underline{}$

3^{10}

and also $(3^7)^2$ as $\underline{}$.

3^{14}

Then $\dfrac{(3^5)^2 \cdot 3^6}{(3^7)^2} = \dfrac{\underline{} \cdot 3^6}{\underline{}} = \dfrac{3^{\underline{}}}{3^{\underline{}}} = \underline{}$.

$\dfrac{3^{10} \cdot 3^6}{3^{14}}$; $\dfrac{3^{16}}{3^{14}}$; 3^2

16. Simplify: $\dfrac{9^2 \cdot 9^5 \cdot (9^2)^3}{9^8} = \dfrac{9^2 \cdot 9^5 \cdot \underline{}}{9^8}$

$\dfrac{9^2 \cdot 9^5 \cdot 9^6}{9^8}$

$= \dfrac{9^{\underline{}}}{9^8} = \underline{}$.

$\dfrac{9^{13}}{9^8}$; 9^5

17. Simplify: $\dfrac{(4^3)^2 \cdot (4^4)^3}{(4^5)^3 \cdot 4} = \dfrac{4^{\underline{}} \cdot 4^{\underline{}}}{4^{\underline{}} \cdot 4^{\underline{}}} = \dfrac{4^{\underline{}}}{4^{\underline{}}}$.

$\dfrac{4^6 \cdot 4^{12}}{4^{15} \cdot 4^1}$; $\dfrac{4^{18}}{4^{16}}$

The exponent on 4 is understood to be $\underline{}$.

1

Final answer: $\underline{}$

4^2

3.5 The Quotient Rule and Integer Exponents

Simplify each of the following.

18. $\dfrac{5^{11} \cdot (5^2)^3}{(5^4)^5 \cdot 5^2} = $ _____ $\quad\dfrac{1}{5^5}$

19. $\dfrac{4^3 \cdot (4^5)^2 \cdot 4^9}{(4^{11})^2} = $ _____ $\quad 1$

20. $\dfrac{6^{-4} \cdot 6^3}{6^{-3}} = $ _____ $\quad \dfrac{6^{-1}}{6^{-3}} = 6^{-1-(-3)} = 6^2$

21. $\dfrac{5^{-3} \cdot 5}{5^4 \cdot 5^{-7}} = $ _____ $\quad \dfrac{5^{-2}}{5^{-3}} = 5^1 = 5$

22. $\dfrac{7^{-4} \cdot 7^6}{7^5 \cdot 7^{-3}} = $ _____ $\quad \dfrac{7^2}{7^2} = 1$

23. $(3^{-4})^3 = $ _____ $\quad 3^{-12} = \dfrac{1}{3^{12}}$

24. $(8^{-4})^{-3} = $ _____ $\quad 8^{12}$

25. $(10^{-1})^{-5} = $ _____ $\quad 10^5$

26. $\left(\dfrac{3x^{-2}}{2y^{-2}}\right)^{-2} = $ _____ $\quad \dfrac{4x^4}{9y^4}$

27. All rules for exponents also work with variables, as long as the variables represent _____ . $\quad$ integers

Simplify each of the following. Assume all variables used as exponents represent integers.

28. $5y^r \cdot (7y^{r+3}) = $ _____ $\quad 35y^{2r+3}$

29. $\dfrac{p^{9q}}{p^{2q}} = $ _____ $\quad p^{7q}$

30. $a^{m+1} \cdot a^{3-m} \cdot a^{5+4m} = $ _____ $\quad a^{9+4m}$

Chapter 3 Polynomials and Exponents

3.6 The Quotient of a Polynomial and a Monomial

[1] Divide a polynomial by a monomial. (See Frames 1–12 below.)

---

1. A monomial is a polynomial of _____ term, such as $4x^3$, $-3m^4$, $-5p^5$, or $2m^2$. To divide the polynomial $12m^5 - 8m^3 + 16m^2$ by the monomial $2m^2$, divide each _____ of the polynomial by _____.

$$\frac{12m^5 - 8m^3 + 16m^2}{2m^2} = \frac{12m^5}{\underline{}} - \frac{8m^3}{2m^2} + \frac{16m^2}{\underline{}}$$

$$= \underline{} - \underline{} + \underline{}$$

one

term; $2m^2$

$2m^2$; $2m^2$
$6m^3$; $4m$; 8

2. $\dfrac{14k^4 - 7k^3 + 21k^2}{7k} = \dfrac{14k^4}{\underline{}} - \dfrac{7k^3}{\underline{}} + \dfrac{21k^2}{7k}$

$= \underline{}$

To check this answer, multiply _____ by _____. The result should be _____.

$7k$; $7k$
$2k^3 - k^2 + 3k$

$2k^3 - k^2 + 3k$
7;
$14k^4 - 7k^3 + 21k^2$

Divide the following. Check your answers by multiplying.

3. $\dfrac{25r^5 - 10r^4 + 50r^3}{5r^3} = \underline{}$

$5r^2 - 2r + 10$

4. $\dfrac{96p^5 - 84p^4 - 72p^3}{12p^2} = \underline{}$

$8p^3 - 7p^2 - 6p$

5. $\dfrac{3p^5 - 9p^3 + 6p^2}{3p^3} = \dfrac{3p^5}{3p^3} - \dfrac{9p^3}{\underline{}} + \dfrac{6p^2}{\underline{}}$

$= \underline{}$

$3p^3$; $3p^3$
$p^2 - 3 + \dfrac{2}{p}$

6. $\dfrac{8y^9 - 16y^5 + 24y^3}{8y^4} = \dfrac{8y^9}{\underline{}} - \dfrac{16y^5}{\underline{}} + \dfrac{24y^3}{\underline{}}$

$= \underline{}$

$8y^4$; $8y^4$; $8y^4$
$y^5 - 2y + \dfrac{3}{y}$

7. $\dfrac{150m^6 - 100m^4 + 80m^3 - 120m^2}{10m^4} = \underline{}$

$15m^2 - 10 + \dfrac{8}{m} - \dfrac{12}{m^2}$

3.7 The Quotient of Two Polynomials

8. $\dfrac{45x^6 - 54x^4 + 63x^3}{9x^6} =$ _____ | $5 - \dfrac{6}{x^2} + \dfrac{7}{x^3}$

9. $\dfrac{20a^4 - 40a^3 + 15a^2 - 5}{10a^2} =$ _____ | $2a^2 - 4a + \dfrac{3}{2} - \dfrac{1}{2a^2}$

10. $\dfrac{18r^{10} - 9r^8 + 27r^6 - 36r^5}{36r^6}$

    $= \dfrac{18r^{10}}{\_\_\_\_} - \dfrac{9r^8}{\_\_\_\_} + \dfrac{27r^6}{\_\_\_\_} - \dfrac{36r^5}{\_\_\_\_}$

    $= _____$

    $36r^6$; $36r^6$; $36r^6$; $36r^6$

    $\dfrac{r^4}{2} - \dfrac{r^2}{4} + \dfrac{3}{4} - \dfrac{1}{r}$

11. $\dfrac{25b^{10} - 40b^8 + 30b^7 + 20b^6}{15b^8} =$ _____ | $\dfrac{5b^2}{3} - \dfrac{8}{3} + \dfrac{2}{b} + \dfrac{4}{3b^2}$

12. $\dfrac{50z^7 - 30z^5 + 20z^3 - 15z}{-20z^4} =$ _____ | $-\dfrac{5z^3}{2} + \dfrac{3z}{2} - \dfrac{1}{z} + \dfrac{3}{4z^3}$

3.7 The Quotient of Two Polynomials

[1] Divide a polynomial by a polynomial. (See Frames 1-8 below.)

---

1. To divide one polynomial, such as $12m^3 - 23m^2 + 13m - 2$, by another, such as $3m - 2$, it is necessary to go through several steps, as indicated below. Note the similarity of these steps to division of _____ numbers. | whole

   Step 1: Write the problem.

   $3m - 2 \overline{)12m^3 _____}$ | $\overline{)12m^3 - 23m^2 + 13m - 2}$

   Step 2: _____ divides into _____ a total of _____ times. Multiply _____ times _____:

   $4m^2(3m - 2) =$ _____.

   $3m - 2 \overline{)12m^3 - 23m^2 + 13m - 2}$

   _____

   $3m$; $12m^3$

   $4m^2$; $4m^2$; $3m - 2$

   $12m^3 - 8m^2$

   $4m^2$

   $12m^3 - 8m^2$

Chapter 3 Polynomials and Exponents

Step 3: _____, and bring down the next term. To subtract, _____ the signs of the _____ polynomial, and _____.

$$3m - 2 \overline{\smash{)}\begin{array}{r} 4m^2 \\ 12m^3 - 23m^2 + 13m - 2 \\ \underline{12m^3 - 8m^2} \end{array}}$$

Subtract
change
bottom; add

$-15m^2 + 13m$

Step 4: _____ divides into _____ a total of _____ times. Multiply:
$-5m(3m - 2) =$ _____.

$$3m - 2 \overline{\smash{)}\begin{array}{r} 4m^2 \\ 12m^3 - 23m^2 + 13m - 2 \\ \underline{12m^3 - 8m^2} \\ -15m^2 + 13m \\ \underline{} \end{array}}$$

$3m; -15m^2$
$-5m$
$-15m^2 + 10m$
$-5m$

$-15m^2 + 10m$

Step 5: Subtract, and _____ down the next term, which is ____.

$$3m - 2 \overline{\smash{)}\begin{array}{r} 4m^2 - 5m \\ 12m^3 - 23m^2 + 13m - 2 \\ \underline{12m^3 - 8m^2} \\ -15m^2 + 13m \\ \underline{-15m^2 + 10m} \\ \underline{} \end{array}}$$

bring
-2
$-5m$

$3m - 2$

Step 6: Divide 3m into ____, a total of ____ times. Complete the problem.

$$3m - 2 \overline{\smash{)}\begin{array}{r} 4m^2 - 5m + \underline{} \\ 12m^3 - 23m^2 + 13m - 2 \\ \underline{12m^3 - 8m^2} \\ -15m^2 + 13m \\ \underline{-15m^2 + 10m} \\ 3m - 2 \\ \underline{} \\ 0 \end{array}}$$

$3m; 1$
1

$3m - 2$

Is there a remainder? (yes/no) The quotient here can be written as:

$$\frac{12m^3 - 23m^2 + 13m - 2}{3m - 2} = \underline{}.$$

no

$4m^2 - 5m + 1$

3.7 The Quotient of Two Polynomials

2. $2x - 3 \overline{)6x^3 - 17x^2 + 22x - 15}$

   0

   $3x^2 - 4x + 5$

   $6x^3 - 9x^2$

   $-8x^2 + 22x$
   $-8x^2 + 12x$

   $10x - 15$
   $10x - 15$

3. $4x - 3 \overline{)16x^3 - 44x^2 + 36x - 9}$

   $4x^2 - 8x + 3$

   $16x^3 - 12x^2$

   $-32x^2 + 36x$
   $-32x^2 + 24x$

   $12x - 9$
   $12x - 9$
   0

4. $3r - 5 \overline{)6r^3 - 34r^2 + 49r - 20}$

   -5

   $2r^2 - 8r + 3$

   $6r^3 - 10r^2$

   $-24r^2 + 49r$
   $-24r^2 + 40r$

   $9r - 20$
   $9r - 15$

   The remainder here is written $\dfrac{-5}{3r - 5}$, so that the answer is not a _____.

   $\dfrac{-5}{3r - 5}$

   polynomial

   $\dfrac{6r^3 - 34r^2 + 49r - 20}{3r - 5} =$ _____

   $2r^2 - 8r + 3 + \dfrac{-5}{3r - 5}$

5. $2m + 3 \overline{)12m^3 + 2m^2 - 18m + 12}$

   Answer: _____

   $6m^2 - 8m + 3 + \dfrac{3}{2m + 3}$

128   Chapter 3 Polynomials and Exponents

6. $3r - 4 \overline{) 12r^3 - 40r^2 + 35r - 8}$

Answer: _____ | $4r^2 - 8r + 1 + \dfrac{-4}{3r - 4}$

7. $2m - 3 \overline{) -16m^4 + 28m^3 - 12m^2 + 17m + 12}$

Answer: $-8m^3 + 2m^2$ _____ | $-3m + 4 + \dfrac{24}{2m - 3}$

8. $x^2 - 3 \overline{) 2x^4 + 4x^3 - 7x^2 - 12x + 3}$ | $2x^2 + 4x - 1$

3.8 An Application of Exponents: Scientific Notation

[1] Express numbers in scientific notation. (See Frames 1-3 below.)

[2] Convert numbers in scientific notation to numbers without exponents. (Frames 4-6)

[3] Use scientific notation in calculations. (Frames 9-15)

---

1. One very important use of exponents is in science, where they are used to write very large or very small numbers. For example,

100 = 10—	10^2
1000 = 10—	10^3
10,000 = 10—	10^4
100,000 = 10—	10^5
1,000,000 = 10—.	10^6

2. The number 5,000,000 can be written
   5,000,000 = (   ) × (1,000,000). | 5
   Since 1,000,000 = 10—, | 10^6
   then 5,000,000 = ____ × 10—. | 5×10^6

3.8 An Application of Exponents: Scientific Notation

Also, $8,000,000,000 = \underline{} \times 10^{\underline{}}$.	8×10^9
This method of writing numbers is called \underline{} notation.	scientific

3. Write each number in scientific notation.

$8000 = \underline{} \times 10^{\underline{}}$	8×10^3
$98,000,000 = 9.8 \times 10^{\underline{}}$	9.8×10^7
$639,000,000 = \underline{} \times 10^{\underline{}}$	6.39×10^8
$15,340,000,000 \underline{} \times 10^{\underline{}}$	1.534×10^{10}
$240,000,000,000 = \underline{}$	2.4×10^{11}
$3,428,000,000 = \underline{}$	3.428×10^9

4. You can also work backwards. To write 9.38×10^{11} without exponents, write down \underline{}. The final number should contain \underline{} digits after the 9. Thus,

	938
	11
$9.38 \times 10^{11} = \underline{}$.	938,000,000,000

Write each number without exponents.

$8.13 \times 10^5 = \underline{}$	813,000
$4.293 \times 10^{12} = \underline{}$	4,293,000,000,000
$5.69 \times 10^4 = \underline{}$	56,900
$9.42 \times 10^0 = \underline{}$	9.42
$2.176 \times 10^5 = \underline{}$	217,600

5. Small numbers can be written in scientific notation by using \underline{} exponents. For example,

	negative
$.0000009 = 9 \times 10^{\underline{}}$.	9×10^{-7}

To find the exponent, \underline{} in this example, count from the right of the first nonzero digit to the \underline{} point. Write each number in scientific notation.

	-7
	decimal
$.000374 = 3.74 \times 10^{\underline{}}$	3.74×10^{-4}
$.0000972 = 9.72 \times 10^{\underline{}}$	9.72×10^{-5}
$.0000000815 = \underline{}$	8.15×10^{-8}
$.938 = \underline{}$	9.38×10^{-1}
$.07 = \underline{}$	7×10^{-2}

Chapter 3 Polynomials and Exponents

6. Write each number without exponents.

   8.6×10^{-5} = _____   .000086

   9.47×10^{-8} = _____   .0000000947

   1.79×10^{-6} = _____   .00000179

   2.5×10^{-3} = _____   .0025

7. Simplify $(2 \times 10^4) \times (4 \times 10^{-6})$.
   Use properties of exponents.

   $(2 \times 10^4) \times (4 \times 10^{-6})$

   = (___ × ___) × (10^4 × ___)   2; 4; 10^{-6}

   = (___) × 10—   8; −2

   = ___   .08

8. $\dfrac{9 \times 10^6}{3 \times 10^{-2}}$ = _____ 3×10^8, or 300,000,000

9. $(2 \times 10^{-4}) \times (3 \times 10^7)$ = _____ 6×10^3, or 6000

10. $\dfrac{(9 \times 10^{-2}) \times (3 \times 10^{-1})}{1 \times 10^{-5}}$ = _____   2700

Chapter 3 Test

The answers for these questions are at the back of this Study Guide.

Evaluate each expression.

1. $\left(\frac{2}{5}\right)^{-2}$

2. $(-7)^{-3}$

Simplify. Write each answer using only positive exponents.

3. $9^{-2} \cdot 9^4 \cdot 9^3 \cdot 9^{-7}$

4. $\dfrac{(2^{-1})^{-2} \cdot 2^{-6}}{2^{-4} \cdot 2^{-3}}$

5. $\left[\dfrac{2z^{-1}}{z^{-3}}\right]^{-2}$

6. $\dfrac{(5r^{-1})^{-2}(5r^3)^2}{(5^{-1}r^{-2})^{-3}}$

7. Write .0000908 in scientific notation.

Write each number without exponents.

8. $(2 \times 10^{-5}) \times (3 \times 10^7)$

9. $\dfrac{9.2 \times 10^{-5}}{4.6 \times 10^{-7}}$

For each polynomial, combine terms; then give the degree of the polynomial. Finally, select the most specific description from this list: (a) trinomial, (b) binomial, (c) monomial, (d) none of these.

10. $9r^3 - 8r + 3r^2$

11. $2p - 5p + 8p^2 - 11p^2 + 7p^2$

Chapter 3 Polynomials and Exponents

12. Subtract.

$$-5m^3 + 2m^2 - 7m + 3$$
$$8m^3 - 5m^2 + 9m - 5$$

12. _____

Perform the indicated operation.

13. $(5m^3 - 9m^2 - 7m + 1) - (3m^3 - 8m^2 - 7m + 2)$ 13. _____

14. $(r^2 - 11r + 7) + (2r^2 - 8r) - (r^2 - 7r + 2)$ 14. _____

15. $7p^3(2p^2 - 4p + 9)$ 15. _____

16. $(3q - 2)(4q + 7)$ 16. _____

17. $(3a + 5b)(4a - 7b)$ 17. _____

18. $(5z + 7)(2z^2 - 7z + 4)$ 18. _____

19. $(4p + 9q)^2$ 19. _____

20. $\left(3z - \frac{5}{3}y\right)^2$ 20. _____

21. $(2m^2 - 7p)(2m^2 + 7p)$ 21. _____

22. $\dfrac{-8x^5 - 16x^4 + 12x^3}{8x^4}$ 22. _____

23. $(21m^3 + 28m^2 - 14m + 7) \div (14m^3)$ 23. _____

24. $\dfrac{6p^3 - p^2 - 29p + 14}{3p + 7}$ 24. _____

25. $\dfrac{15r^4 - 24r^3 + 31r^2 - 16r + 10}{3r^2 + 2}$ 25. _____

CHAPTER 4 FACTORING AND APPLICATIONS

4.1 Factors; The Greatest Common Factor

[1] Find the greatest common factor of a list of terms. (See Frames 1-6 below.)

[2] Factor out the greatest common factor. (Frames 7-14)

[3] Factor by grouping. (Frames 15-20)

---

1. In Chapter 1 you wrote numbers as products of prime _____. A factor of two or more terms is called a _____ factor for the two terms. For example, 8 is a common factor of 16 and 24, since 8 _____ into both _____ and 24. Another example: 12 (*is/is not*) a common factor of 36 and 72, since 12 _____ into both 36 and 72 without a remainder.

   factors
   common

   divides; 16
   is
   divides

2. $3x$ (*is/is not*) a common factor of $12x^2$ and $9x$.

   is

3. $8m^2$ (*is/is not*) a common factor of $16m^4$ and $32m$. This is because $8m^2$ (*is/is not*) a factor of $32m$.

   is not
   is not

4. The _____ of the common factors of two or more terms is called the _____ common factor. The greatest common factor for 12 and 18 is ___, since 6 is the largest integer factor of both 12 and 18. Find the greatest common factors for the following terms.

   greatest
   greatest
   6

24, 32, 40	8
50, 60, 80	10
$3x$, $9x$, $12x$	$3x$
$8m^2$, $4m$, $12m^2$	$4m$
$14p^3$, $21p^4$, $35p^5$	$7p^3$

Chapter 4 Factoring and Applications

When looking for the greatest common factor for a series of terms involving powers of the same variable, use the _____ exponent from any of the powers.

smallest

Find the greatest common factor in Frames 5 and 6.

5. $30r^7$, $15r^6$, $45r^9$

   The greatest common factor for 30, 15, and 45 is ____. The greatest common factor for r^7, r^6, and r^9 is ____. (Use the _____ exponent.) Therefore, the greatest common factor for $30r^7$, $15r^6$, and $45r^9$ is ____.

15

r^6; smallest

$15r^6$

6. $8m^3n^4$, $12m^2n^5$, $16m^3n^8$, $20m^9n^3$

   Greatest common factor: _____

$4m^2n^3$

7. The greatest common factor of the terms of the polynomial $12p^4 + 8p^5$ is ____. Rewrite these terms using the greatest common factor.

   $12p^4 + 8p^5 = (4p^4)(\ \ ) + (4p^4)(\ \ )$

   Using the _____ property, we have

   $12p^4 + 8p^5 = (\ \ )(\ \ )$.

   To check this result, multiply ____ and ____. The result should be _____.

$4p^4$

3; 2p

distributive

$4p^4$; $3 + 2p$

$4p^4$; $3 + 2p$
$12p^4 + 8p^5$

8. In Frame 7, you wrote $12p^4 + 8p^5$ as $4p^4(3 + 2p)$. This process is called _____ out the _____ common factor.

factoring
greatest

4.1 Factors; The Greatest Common Factor

Factor out the greatest common factor in Frames 9–14.

9. $25m^4 + 30m^3$

   The greatest common factor is _____. | $5m^3$

   $25m^4 + 30m^3 = (5m^3)() + (5m^3)()$ | $5m;\ 6$

   $ = ()()$ | $5m^3;\ 5m + 6$

10. $36r^3 - 12r^2 + 24r = ()()$ | $12r;\ 3r^2 - r + 2$

11. $36m - 18 = ()()$ | $18;\ 2m - 1$

12. $9p^5 - 18p^4 + 36p^3 = ()()$ | $9p^3;\ p^2 - 2p + 4$

13. $12y^4x^2 - 18y^2x^3 = ()()$ | $6y^2x^2;\ 2y^2 - 3x$

14. $30p^4q^3 - 42pq^4 = ()()$ | $6pq^3;\ 5p^3 - 7q$

15. Common factors are used in factoring by _____, where terms are grouped two or more at a time. | grouping

Factor by grouping in Frames 16–20.

16. $4m + 12 + mp + 3p = 4() + p()$ | $m + 3;\ m + 3$

    $ = \underline{}$ | $(m + 3)(4 + p)$

17. $q^2 + 9q + rq + 9r = \underline{}$ | $(q + 9)(q + r)$

18. $z^2 - 4z + 5z - 20 = z() + 5()$ | $z - 4;\ z - 4$

    $ = \underline{}$ | $(z - 4)(z + 5)$

19. $3a^2 + 7a + 6a + 14 = \underline{}$ | $(3a + 7)(a + 2)$

20. $8x^2 - 10x + 4x - 5 = \underline{}$ | $(4x - 5)(2x + 1)$

Chapter 4 Factoring and Applications

4.2 Factoring Trinomials

[1] Factor trinomials with a coefficient of 1 for the squared term. (See Frames 1–18 below.)

[2] Factor such polynomials after factoring out the greatest common factor. (Frames 19–22)

---

1 An expression is written in _____ form when it is written as the product of two or more _____. In this section, you will factor trinomials like $x^2 - 5x + 6$, where the coefficient of the x^2 term is ___.	factored factors 1
2. To factor $x^2 - 5x + 6$, first list all pairs of integers whose product is ____.	6
$\qquad 6 \cdot \underline{}$	1
$\qquad 3 \cdot \underline{}$	2
$\qquad -6 \cdot \underline{}$	-1
$\qquad -3 \cdot \underline{}$	-2
Now select from this list the pair of integers whose sum is ____. This pair of integers is ____ and ____, so that $x^2 - 5x + 6$ factors as	-5 -3; -2
$\qquad x^2 - 5x + 6 = ()()$.	$(x - 3)(x - 2)$

Factor each trinomial.

3. $x^2 + 2x - 15$. List all pairs of integers whose product is ____.	-15
$\qquad 15 \cdot \underline{}$	-1
$\qquad 5 \cdot \underline{}$	-3
$\qquad -5 \cdot \underline{}$	3
$\qquad -15 \cdot \underline{}$	1
The pair of integers whose sum is ____ is ____ and ____. Therefore, $x^2 + 2x - 15 = ()()$. This answer can also be written as $()()$.	2; 5 -3; $x + 5$; $x - 3$ $x - 3$; $x + 5$

4.2 Factoring Trinomials   137

4. $r^2 + 9r + 20$
    The two numbers whose product is ____ and whose sum is ____ are ____ and ____. Therefore,
    $r^2 + 9r + 20 = ($ $)($ $).$

    20
    9; 4; 5
    r + 4; r + 5

5. $m^2 - 7m + 12$
    The two numbers whose product is ____ and whose sum is ____ are ____ and ____. Therefore,
    $m^2 - 7m + 12 = ($ $)($ $).$

    12
    -7; -3; -4
    m - 3; m - 4

6. $y^2 + y - 30$
    Here, the two numbers we need are ____ and ____, so that
    $y^2 + y - 30 = ($ $)($ $).$

    6; -5

    y + 6; y - 5

7. $k^2 + 7k + 8$
    We cannot find two integers whose sum is ____ and whose product is ____. This polynomial (*can/cannot*) be factored.

    7
    8
    cannot

8. A polynomial that cannot be factored is called a _____ polynomial.

    prime

Factor in Frames 9–22.

9. $a^2 + 10a + 21 =$ _____ $(a + 3)(a + 7)$

10. $p^2 + 5p - 14 =$ _____ $(p + 7)(p - 2)$

11. $m^2 - 5m - 24 =$ _____ $(m + 3)(m - 8)$

12. $d^2 - 2d - 35 =$ _____ $(d + 5)(d - 7)$

13. $r^2 - r - 72 =$ _____ $(r + 8)(r - 9)$

Chapter 4 Factoring and Applications

14. $y^2 - 4y - 45 =$ _____ $(y + 5)(y - 9)$

15. $n^2 + 20n + 99 =$ _____ $(n + 9)(n + 11)$

16. $m^2 - 8mn + 12n^2$

    We need two terms whose product is _____ and whose sum is _____. These terms are _____ and _____. Factor as

    $m^2 - 8mn + 12n^2 =$ _____.

    $12n^2$
    $-8n; -6n$
    $-2n$

    $(m - 6n)(m - 2n)$

17. $p^2 + 9pq + 8q^2 =$ _____ $(p + 8q)(p + q)$

18. $z^2 + 15zx + 50x^2 =$ _____ $(z + 10x)(z + 5x)$

19. $3a^3 + 3a^2 - 60a$

    This trinomial has a greatest common factor of ____. Factoring out $3a$ gives

    $3a^3 + 3a^2 - 60a =$ _____.

    Factor $a^2 + a - 20$, so that the final factored form is

    $3a^3 + 3a^2 - 60a =$ _____

    $3a$

    $3a(a^2 + a - 20)$

    $3a(a + 5)(a - 4)$

20. $5r^5 + 20r^4 - 60r^3 = ($ $)($ $)$
    $= ($ $)($ $)($ $)$

    $5r^3; r^2 + 4r - 12$
    $5r^3; r + 6; r - 2$

21. $2y^6 + 8y^5 - 64y^4 =$ _____ $2y^4(y - 4)(y + 8)$

22. $8z^4m - 40z^3m - 48z^2m =$ _____ $8z^2m(z - 6)(z + 1)$

4.3 More on Factoring Trinomials

☐1 Factor trinomials by grouping when the coefficient of the squared term is not 1. (See Frames 1–9)

☐2 Factor trinomials by trial and error. (Frames 10–27)

---

1. Now we can factor trinomials where the coefficient of the squared term is not ___.

   1

2. Recall that a trinomial such as $x^2 + 4x - 21$ is factored by finding two numbers whose product is ____ and whose sum is ___.

   −21; 4

3. To factor $2x^2 + 9x + 7$, look for two numbers whose product is $2 \cdot 7 =$ ____, and whose sum is ___.

   14
   9

   ┌──── Sum is 9
   $2x^2 + 9x + 7$
   └──────┘
   Product is _____ = ____

   2 · 7; 14

   The numbers having a product of 14 and a sum of 9 are ___ and ___. Use these numbers and write the middle term, 9x, as _____. This gives.

   7; 2
   7x + 2x

   $2x^2 + 9x + 7 = 2x^2 +$ _____ $+ 7$,
   which we now factor by _____:
   = x(   ) + ____ (2x + 7)
   = _____.

   7x + 2x
   grouping
   2x + 7; 1
   (2x + 7)(x + 1)

Factor each trinomial in Frames 4–9.

4. $8m^2 + 26m + 15$
   Find two numbers whose product is _____ and whose sum is ____. The numbers are ____ and ____. Write 26m as _____.

   8 · 15 = 120
   26; 20
   6; 20m + 6m

140   Chapter 4   Factoring and Applications

$$8m^2 + 26m + 15 = 8m^2 + 20m + 6m + 15$$
$$= 4m(2m + 5) + \underline{\qquad}$$ $3(2m + 5)$
$$= \underline{\qquad\qquad}$$ $(2m + 5)(4m + 3)$

5. $6k^2 - 11k + 3$

    Find two numbers having a product of _____ and $6 \cdot 3 = 18$
    a sum of ____. The numbers are ____ are ____. $-11; -9; -2$
    Write $-11k$ as _____. Now $-9k + (-2k)$

$$6k^2 - 11k + 3 = 6k^2 - 9k + (-2k) + 3$$
$$= \underline{\qquad} + \underline{\qquad}$$ $3k(2k - 3);$
    $-1(2k - 3)$
$$= \underline{\qquad\qquad}.$$ $(2k - 3)(3k - 1)$

6. $5r^2 + 22r + 8 = \underline{\qquad\qquad}$ $(5r + 2)(r + 4)$

7. $2y^2 - y - 15 = \underline{\qquad\qquad}$ $(2y + 5)(y - 3)$

8. $12m^2 - m - 35 = \underline{\qquad\qquad}$ $(4m - 7)(3m + 5)$

9. $25a^2 + 5a - 2 = \underline{\qquad\qquad}$ $(5a + 2)(5a - 1)$

10. As a method of factoring, we can list possible
     _____ and see which ones work. This is   factors
     called factoring by _____.   trial and error

11. To factor $2x^2 + 5x - 3$, first list the possible
     factors of $2x^2$: ____ and ____. ($-2x$ and $-x$ are $2x; x$
     also factors, but normally it is not necessary to
     consider negative factors for the first term.)
     Now list the possible factors of -3.

     ____ and ____ or ____ and ____ $3; -1; -3; 1$

     Now list all possible combinations of factors of
     $2x^2$ and -3, and keep checking until you get the
     desired product, which is _____. $2x^2 + 5x - 3$

4.3 More on Factoring Trinomials

Possible factors	Correct?	
$(2x + 3)(x - 1) =$ _____	_____	$2x^2 + x - 3$; no
$(2x - 3)(x + 1) =$ _____	_____	$2x^2 - x - 3$; no
$(2x + 1)(x - 3) =$ _____	_____	$2x^2 - 5x - 3$; no
$(2x - 1)(x + 3) =$ _____	_____	$2x^2 + 5x - 3$; yes

Therefore, $2x^2 + 5x - 3 =$ _____. $(2x - 1)(x + 3)$

Factor each trinomial.

12. $3r^2 + 10r - 8$

    Possible factors of $3r^2$ are ____ and ____. $3r$; r
    Possible factors of -8 are

    $-8 \cdot$ _____   1
    $-4 \cdot$ _____   2
    $-1 \cdot$ _____   8
    $4 \cdot$ _____. -2

    Try the various possibilities and find the factors of $3r^2 + 10r - 8$.

    $3r^2 + 10r - 8 =$ _____ $(3r - 2)(r + 4)$

13. $5y^2 + 13y + 6$

    Factors of $5y^2$ are ____ and ____. Factors of 6 $5y$; y
    are

    $6 \cdot$ _____   1
    $3 \cdot$ _____   2
    $-3 \cdot$ _____ -2
    $-6 \cdot$ _____ -1

    Since the middle term of the expression you are factoring, _____, has a + sign, you would not $13y$
    need to use factors having _____ signs.   minus
    Complete the factoring.

    $5y^2 + 13y + 6 =$ _____ $(5y + 3)(y + 2)$

Chapter 4 Factoring and Applications

14. $6y^2 - y - 12$

    Factors of $6y^2$ are

    $6y \cdot$ ____   y

    $3y \cdot$ ____ .   2y

    It is often a good idea to start with the factors of moderate size, which are ____ and ____ here.   3y; 2y
    Factors of -12 are

    $12 \cdot$ ____ -1

    $6 \cdot$ ____ -2

    $4 \cdot$ ____ -3

    $3 \cdot$ ____ -4

    $2 \cdot$ ____ -6

    $1 \cdot$ ____ -12

    It might be a good idea to start with 4 and ____, -3
    or -4 and ____. Complete the factoring.   3

    $6y^2 - y - 12 =$ _____ $(2y - 3)(3y + 4)$

15. $3x^2 + x + 2$.

    The factors of $3x^2$ are ____ and ____. The factors of 2 that we would want to try here (because the middle term has a ____ sign) are ____ are ____.   3x; x

    The possible factors are given as follows.   +; 2; 1

                Correct?

    $(3x + 2)(x + 1) =$ _____ ____ $3x^2 + 5x + 2$; no

    $(3x + 1)(x + 2) =$ _____ ____ $3x^2 + 7x + 2$; no

    There are no factors that work here, so that $3x^2 + x + 2$ (*can/cannot*) be factored with integer factors. This polynomial is _____.

    cannot

    prime

Factor by either method in Frames 16–27.

16. $6r^2 + 11r - 10 =$ _____ $(3r - 2)(2r + 5)$

17. $2k^2 + k - 21 =$ _____ $(2k + 7)(k - 3)$

18. $9m^2 + 9m - 10 =$ _____ $(3m + 5)(3m - 2)$

19. $15r^2 + 7r - 2 =$ _____ $(3r + 2)(5r - 1)$

20. $6p^2 + 7p - 5 =$ _____ $(3p + 5)(2p - 1)$

21. $12r^2 + 11r - 15 =$ _____ $(4r - 3)(3r + 5)$

22. $6x^2 + 11xy - 7y^2 =$ _____ $(3x + 7y)(2x - y)$

23. $12g^2 - 4gm - 5m^2 =$ _____ $(6g - 5m)(2g + m)$

24. To factor $24m^3 - 42m^2 + 9m$, first factor out the _____ factor, giving

    common

    $24m^3 - 42m^2 + 9m = ($ $)($ $)$

    $3m;\ 8m^2 - 14m + 3$

    Complete the factoring.

    $24m^3 - 42m^2 + 9m =$ _____

    $3m(2m - 3)(4m - 1)$

25. $8r^3s + 8r^2s^2 - 30rs^3 = ($ $)($ $)$

    $= $ _____

    $2rs;$
    $4r^2 + 4rs - 15s^2$
    $2rs(2r-3s)(2r+5s)$

26. $24m^5 + 4m^4n - 60m^3n^2 = ($ $)($ $)$

    $= $ _____

    $4m^3;$
    $6m^2 + mn - 15n^2$
    $4m^3(3m+5n)(2m-3n)$

27. $30x^3y^2 - 57x^2y^3 - 45xy^4 =$ _____ $3xy^2(5x+3y)(2x-5y)$

4.4 Special Factorizations

[1] Factor the difference of two squares. (See Frames 1-8 below.)

[2] Factor a perfect square trinomial. (Frames 9-18)

[3] Factor the difference of two cubes. (Frames 19-22)

[4] Factor the sum of two cubes. (Frames 23-26)

144   Chapter 4 Factoring and Applications

1. To factor $x^2 - y^2$, write $x^2 - y^2 =$ (   )(   ). | $x + y; x - y$
   This factorization is called the _____ of | difference
   two squares. For example,

   $x^2 - 16 = x^2 -$ (   )$^2 =$ _____. | $4; (x + 4)(x - 4)$

Factor each polynomial in Frames 2–7.

2. $9y^2 - 4 =$ (   )$^2 -$ (   )^2 | $3y; 2$
   $=$ _____ | $(3y + 2)(3y - 2)$

3. $16p^2 - 25 =$ (   )$^2 -$ (   )^2 | $4p; 5$
   $=$ _____ | $(4p + 5)(4p - 5)$

4. $121k^2 - 25 =$ _____ | $(11k + 5)(11k - 5)$

5. $45m^2 - 20 =$ (   )(   ) | $5; 9m^2 - 4$
   $= 5$ _____ | $(3m + 2)(3m - 2)$

6. $64r^2 - 100 =$ (   )(   ) | $4; 16r^2 - 25$
   $=$ _____ | $(4)(4r + 5)(4r - 5)$

7. $m^4 - n^4 =$ (   )(   ) | $m^2 + n^2; m^2 - n^2$
   $= (m^2 + n^2)$ _____ | $(m + n)(m - n)$

8. $x^2 + 25$ (*can/cannot*) be factored. In general, | cannot
   the _____ of two squares cannot be _____. | sum; factored

9. The polynomial $x^2 + 10x + 25$ can be factored as
   $x^2 + 10x + 25 =$ (   )(   ) $=$ (   )^2. | $x + 5; x + 5;$
   | $x + 5$
   Since $x^2 + 10x + 25$ is a trinomial which equals
   the square of a _____, then $x^2 + 10x + 25$ is | binomial
   called a _____ _____ trinomial. The | perfect; square
   middle term of a perfect square trinomial must be
   _____ the product of the two terms from the | twice
   binomial.

4.4 Special Factorizations

10. To see whether or not $9x^2 - 30x + 25$ is a perfect square trinomial, first check that

    $9x^2 = (\quad)^2$ and $25 = (\quad)^2$. | $3x$; 5

    To be a perfect square trinomial, the _____ | middle
    term of $9x^2 - 30x + 25$ must be twice the product
    of ____ and ____. | $3x$; 5

    $30x = 2(\quad)(\quad)$ (*true/false*) | $3x$; 5; true

    Therefore, $9x^2 - 30x + 25$ (*is/is not*) a perfect square trinomial. | is

    $9x^2 - 30x + 25 = (\quad)^2$ | $3x - 5$

    (The middle sign of the factor is the same as the _____ sign of the given trinomial.) | middle

Factor each polynomial in Frames 11–17.

11. $16x^2 - 24x + 9 = (\quad)^2$ | $4x - 3$

12. $9m^2 + 12m + 4 = (\quad)^2$ | $3m + 2$

13. $25y^2 - 30y + 9 = $ _____ | $(5y - 3)^2$

14. $16y^2 + 56y + 49 = $ _____ | $(4y + 7)^2$

15. $4p^2 + 4p + 1 = $ _____ | $(2p + 1)^2$

16. $4a^2 - 20a + 25 = $ _____ | $(2a - 5)^2$

17. $16p^2 - 72p + 81 = $ _____ | $(4p - 9)^2$

18. To factor $32m^2 + 80m + 50$, first factor out the _____ common factor. | greatest

    $32m^2 + 80m + 50 = (\quad)(\quad)$ | 2; $16m^2 + 40m + 25$

Chapter 4 Factoring and Applications

Complete the factoring.

$$32m^2 + 80m + 50 = (\quad)(\quad)^2$$

2; 4m + 5

19. The difference of two _____ is factored with the pattern

$$x^3 - y^3 = (\quad)(\quad).$$

cubes

x − y;
$x^2 + xy + y^2$

Factor in Frames 20–22.

20. $\quad p^3 - 216 = p^3 - (\quad)^3$

$\qquad = (p - 6)(p^2 + \_\_\_ + \_\_\_)$

$\qquad = (p - 6)(\quad)$

6

6p; 6^2

$p^2 + 6p + 36$

21. $\quad 27y^3 - 8z^3 = (\quad)^3 - (\quad)^3$

$\qquad = (3y - 2z)[(\quad)^2 + (\quad)(\quad) + (2z)^2]$

$\qquad = (3y - 2z)(\quad)$

3y; 2z

3y; 3y; 2z

$9y^2 + 6yz + 4z^2$

22. $\quad 24x^3 - 3y^3 = 3(\quad)$

$\qquad = 3(2x - y)(\quad)$

$8x^3 - y^3$

$4x^2 + 2xy + y^2$

23. Factor the _____ of two cubes as

$$x^3 + y^3 = (x + y)(\quad).$$

sum

$x^2 - xy + y^2$

Factor in Frames 24–26.

24. $\quad 64r^3 + 1 = (\quad)^3 + 1^3$

$\qquad = (4r + 1)[(\quad)^2 - (\quad)(\quad) + (\quad)^2]$

$\qquad = (4r + 1)(\quad)$

4r

4r; 4r; 1; 1

$16r^2 - 4r + 1$

25. $\quad 125a^3 + 216b^3 = (\quad)^3 + (\quad)^3$

$\qquad = (5a + 6b)(\quad)$

5a; 6b

$25a^2 - 30ab + 36b^2$

26. $\quad 16z^3 + 54x^3 = 2(\quad)$

$\qquad = 2(2z + 3x)(\quad)$

$8z^3 + 27x^3$

$4z^2 - 6zx + 9x^2$

4.5 Solving Quadratic Equations by Factoring

[1] Solve quadratic equations by factoring. (See Frames 1-19 below.)

[2] Solve other equations by factoring. (Frames 20-22)

---

1. An equation having 2 as the highest exponent on any variable is called a _____ equation. | quadratic
   For example, $2x^2 - 4x + 8 = 0$, $3x^2 = 16$, and $5x^2 - 2x = 0$ are all examples of _____ equations. | quadratic

2. The solution of quadratic equations by factoring depends on the _____ factor property: if the _____ of two numbers is zero, then at least one of the numbers is ___. For example, if $3a = 0$, then we must conclude that ___ = 0. | zero
product
0
a

3. Given the equation $(x - 5)(x + 3) = 0$, we can use the zero-factor property to write _____ = 0 or _____ = 0. Solve these two equations, obtaining x = ___ or x = ___. | x - 5
x + 3
5; -3

4. Solve the equation $(2x - 3)(x + 6) = 0$. By the zero-factor property,

   _____ = 0   or   _____ = 0 | 2x - 3; x + 6
        2x = ___   or   x = ___ | 3; -6
        x = ___   or   x = -6. | $\frac{3}{2}$

Solve the quadratic equations in Frames 5-19.

5. $x^2 + 2x - 3 = 0$
   First factor $x^2 + 2x - 3$ as (   )(   ) = 0. | x - 1; x + 3
   Then

   _____ = 0   or   _____ = 0 | x - 1; x + 3
       x = ___   or   x = 0 ___. | 1; -3

Chapter 4 Factoring and Applications

To check, substitute ___ or ___ for x in the original equation. | 1; −3

If x = 1, If x = −3,

(   )² + 2(   ) − 3 = 0   (   )² + 2(   ) − 3 = 0 | 1; 1; −3; −3

___ = 0   ___ = 0 | 0; 0

(*true/false*)   (*true/false*) | true; true

6. $x^2 + 3x - 10 = 0$

    (   )(   ) = 0 | x + 5; x − 2

    _____ = 0   or   _____ = 0 | x + 5; x − 2

    x = ___   or   x = ___ | −5; 2

7. $x^2 + 2x - 8 = 0$

    (   )(   ) = 0 | x + 4; x − 2

    _____ = 0   or   _____ = 0 | x + 4; x − 2

    x = ___   or   x = ___ | −4; 2

8. $2x^2 - 5x - 3 = 0$

    (   )(   ) = 0 | 2x + 1; x − 3

    _____ = 0   or   _____ = 0 | 2x + 1; x − 3

    x = ___   or   x = ___ | $-\frac{1}{2}$; 3

9. $3x^2 + 4x - 4 = 0$

    (   )(   ) = 0 | 3x − 2; x + 2

    x = ___   or   x = ___ | $\frac{2}{3}$; −2

10. $4x^2 + 3x = 1$

Rewrite the equation as _____ = 0. | $4x^2 + 3x - 1$

Now factor.

    _____ = 0 | (4x − 1)(x + 1)

    x = ___   or   x = ___ | $\frac{1}{4}$; −1

4.5 Solving Quadratic Equations by Factoring

11. $6x^2 - x + 2$

    $x = $ _____ or $x = $ _____ $\dfrac{2}{3}; -\dfrac{1}{2}$

12. $10x^2 + x = 2$

    $x = $ _____ or $x = $ _____ $-\dfrac{1}{2}; \dfrac{2}{5}$

13. $12x^2 + x = 6$

    $x = $ _____ or $x = $ _____ $\dfrac{2}{3}; -\dfrac{3}{4}$

14. To solve $x^2 + x/4 = 15/8$, first multiply through by ____, to clear the fractions.   8

    $(\quad)x^2 + (\quad)\dfrac{x}{4} = (\quad)\dfrac{15}{8}$ 8; 8; 8

    _____ = ____ $8x^2 + 2x$; 15

    _____ = 0 $8x^2 + 2x - 15$

    $(\quad)(\quad) = 0$ $4x - 5$; $2x + 3$

    $x = $ _____ or $x = $ _____ $\dfrac{5}{4}; -\dfrac{3}{2}$

15. $\dfrac{x^2}{4} - \dfrac{37x}{60} + \dfrac{1}{3} = 0$

    Multiply through by _____. Complete the solution.   60

    $x = $ _____ or $x = $ _____ $\dfrac{4}{5}; \dfrac{5}{3}$

16. To solve $9x^2 - 25 = 0$, first factor.

    $(\quad)(\quad) = 0$ $3x + 5$; $3x - 5$

    $x = $ _____ or $x = $ _____ $-\dfrac{5}{3}; \dfrac{5}{3}$

17. Solve $121x^2 = 25$.

    $x = $ _____ or $x = $ _____ $\dfrac{5}{11}; -\dfrac{5}{11}$

Chapter 4 Factoring and Applications

18. Solve $4x^2 + 20x + 25 = 0$. First factor.

    ( )( ) = 0 | $2x + 5$; $2x + 5$

    Since these two factors are the same, there is only _____ answer here: x = _____. | one; $-\frac{5}{2}$

19. Solve $9x^2 - 12x + 4 = 0$.

    x = _____ | $\frac{2}{3}$

20. To solve $(3m - 5)(m^2 - 4m + 3) = 0$, first factor.

    $(3m - 5)(m^2 - 4m + 3) = (3m - 5)($ _____ $)($ _____ $)$ | $m - 3$; $m - 1$

    Place each factor equal to ___. | 0

    $3m - 5 = 0$, _____ = 0, or _____ = 0 | $m - 3$; $m - 1$

    Solve each equation.

    m = _____ or m = _____ or m = _____ | $\frac{5}{3}$; 3; 1

Solve each equation.

21. $(5k - 1)(2k^2 + 3k - 2) = 0$

    k = _____ or k = _____ or k = _____ | $\frac{1}{5}$; $\frac{1}{2}$; -2

22. $z^3 - 16z = 0$

    z = _____ or z = _____ or z = _____ | 0; 4; -4

4.6 Applications of Quadratic Equations

[1] Solve problems about geometric figures. (See Frames 1-6 below.)

[2] Solve problems using the Pythagorean formula. (Frames 7-9)

---

1. The length of a rectangle is 6 more than the width. The area of the rectangle is 27. Find The length and width of the rectangle.

4.6 Applications of Quadratic Equations

Let x = the width of the rectangle and _____ = the length of the rectangle.

"The area is 27" is written as an equation

_____, or _____.

$$x^2 + 6x - 27 = \_\_\_\_$$

$$(\quad)(\quad) = 0$$

$$x = \_\_\_\_ \quad \text{or} \quad x = \_\_\_\_$$

A rectangle cannot have a side of ____, so the only valid answer is x = ____. Thus, the rectangle is ____ by 3 · 3 = ____.

	x + 6
	x(x + 6) = 27; $x^2 + 6x = 27$
	0
	x + 9; x − 3
	−9; 3
	−9
	3
	3; 9

2. The length of a rectangle is twice the width, and the area is 32 square centimeters. Find the width of the rectangle.

   Let x = the width of the rectangle, and ____ = the length. The area is the product of the length and the _____, or

   $$\text{area} = \_\_\_\_ \cdot \_\_\_\_ = \_\_\_\_.$$

   Write "the area is 32" as an equation.

   _____

   Solve this equation to find that the width is _____ centimeters. (The length is _____ centimeters.

	2x
	width
	2x; x; $2x^2$
	$2x^2 = 32$
	4; 8

3. The length of a rectangle is 1 more than twice the width. The area of the rectangle is 36 square centimeters. Find the width of the rectangle.

   _____ centimeters | 4

4. One side of a box is 6 inches long. Another side is 4 more than the third side. The volume of the box is 126. Find the sides of the box.

152 Chapter 4 Factoring and Applications

Let x = the third side, then _____ = the second side. The formula for the volume of a box is	x + 4
V = _____. Using the values of this problem, we have the equation	LWH
_____ = 6(x)(_____).	126; x + 4
Solve this equation.	
The box is 6 inches by _____ inches by _____ inches.	3; 7

5. The perimeter of a figure gives the _____ around the figure. Use this fact in the word problem of Frame 6. — distance

6. The length of a rectangle is 2 inches longer than the width. The area of the rectangle is numerically 11 more than the perimeter. Find the width of the rectangle.

Let x = the width of the rectangle, with _____ = the length. The area of the rectangle is	x + 2
area = _____;	x(x + 2)
while the perimeter, P = _____, is	2L + 2W
perimeter = _____.	2(x + 2) + 2x
Write "the area is numerically 11 more than the perimeter" as an equation.	
x(x + 2) = _____ + 2(x + 2) + 2x	11
Simplify this equation.	
_____ = 11 + _____ + 2x,	$x^2 + 2x$; 2x + 4
or $x^2 - 2x -$ _____ = 0	15
Factor to get	
( )( ) = 0,	x − 5; x + 3
with	
x = _____ or x = _____.	5; −3

4.6 Applications of Quadratic Equations

Reject ____, giving the width of the rectangle as ____ inches. (The length is ____ inches.)	−3 5; 7
7. Recall the Pythagorean formula. If a and b are the lengths of the legs of a right triangle, (a triangle with a 90° angle) and c is the length of the hypotenuse, then _____. 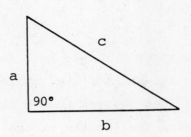	$a^2 + b^2 = c^2$
8. The hypotenuse of a right triangle is 1 inch longer than twice the shorter leg, while the longer leg is 1 inch shorter than twice the shorter leg. Find the lengths of the three sides of the triangle. Let x = the length of the shorter leg. Then _____ = the length of the hypotenuse, and _____ = the length of the longer leg. We draw a right triangle and label it.	$2x + 1$ $2x - 1$
	$2x + 1$ x $2x - 1$
Use the Pythagorean formula to write an equation. $\qquad x^2 + (\quad)^2 = (\quad)^2$	$2x - 1$; $2x + 1$
Simplify. $\qquad x^2 + \underline{\qquad} = 4x^2 + 4x + 1,$ or $\qquad x^2 - 8x = 0.$	$4x^2 - 4x + 1$
Factor: $\underline{\qquad} = 0,$ from which $x = \underline{\quad}$ or $x = \underline{\quad},$	$x(x - 8)$ 0; 8

154   Chapter 4 Factoring and Applications

Reject _____, giving _____ inches as the length | 0; 8
of the shorter leg. The longer leg is 2x − 1 =
_____ inches, and the hypotenuse is 2x + 1 = _____ | 15; 17
inches.

9. The longer leg of a right triangle is 3 feet longer than 3 times the shorter leg, while the hypotenuse is 3 feet shorter than 4 times the shorter leg. Find the lengths of the sides of the triangle.

_____ feet; _____ feet; _____ feet | 7; 24; 25

4.7 Solving Quadratic Inequalities

1 Solve quadratic inequalities and graph their solutions. (See Frames 1−6 below.)

1. So far in this chapter, we have discussed only quadratic equations, such as $x^2 + x - 6 = 0$. In this section we discuss quadratic _____, | inequalities
such as

$$x^2 + x - 6 \geq 0.$$

To solve this inequality, first solve the equation

_____ = 0. | $x^2 + x - 6$

Factor.

( )( ) = 0 | $x + 3$; $x - 2$
x = _____ or x = _____ | −3; 2

These two numbers can be used to divide a number line into three regions.

Region A | Region B | Region C
                0

___ | −3; 2

4.7 Solving Quadratic Inequalities

To solve the original inequality $x^2 + x - 6 \geq 0$, select one point from each of Regions A, B, and C. If the selected point satisfies the inequality, then every point of the region satisfies the inequality. Let us select $x = -4$ from region A. Replace x with -4.

$$x^2 + x - 6 \geq 0$$
$$(\quad)^2 + (\quad) - 6 \geq 0 \qquad \text{−4; −4}$$
$$\underline{\quad} + \underline{\quad} - \underline{\quad} \geq 0 \qquad \text{16; (−4); 6}$$
$$\underline{\quad} \geq 0 \quad (true/false) \qquad \text{6; true}$$

Therefore, _____ point of Region A satisfies the inequality. Now select a point from Region B. Let us select $x = 0$. Then   every

$$x^2 + x - 6 \geq 0$$
$$(\quad)^2 + (\quad) - 6 \geq 0 \qquad \text{0; 0}$$
$$\underline{\quad} \geq 6. \quad (true/false) \qquad \text{−6; false}$$

Therefore, ____ point of Region B satisfies the inequality. Try $x = 3$ from Region C.   no

$$x^2 + x - 6 \geq 0$$
$$(\quad)^2 + (\quad) - 6 \geq 0 \qquad \text{3; 3}$$
$$\underline{\quad} \geq 0 \quad (true/false) \qquad \text{6; true}$$

Therefore, _____ point of Region C satisfies the inequality. In summary, Regions ____ and ____ satisfy the inequality. The solution is written   every

A; C

_____ or _____. $x \leq -3$ or $x \geq 2$

Graph the solution.

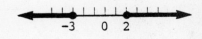

Is there a shortcut way to write $x \leq -3$ or $x \geq 2$? (yes/no)   no

156 Chapter 4 Factoring and Applications

2. Solve $x^2 - 3x - 4 < 0$.

First solve _____ = 0. | $x^2 - 3x - 4$
( )( ) = 0 | $(x - 4)(x + 1)$
x = ____ or x = ____ | 4; -1

Divide the number line into three regions.

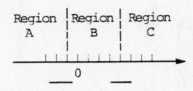

                                         | -1; 4

Try x = -2 from Region A.

$x^2 - 3 - 4 < 0$
( )² - 3( ) - 4 < 0 | -2; -2
_____ < 0 (true/false) | 6; false

Therefore, ____ point from Region A satisfies the | no
inequality. Try x = 0 from Region B.

( )² - 3( ) - 4 < 0 | 0; 0
____ < 0 (true/false) | -4; true

Therefore, all points from Region ____ will | B
satisfy the inequality. Try x = 5 from Region C.

( )² - 3( ) - 4 < 0 | 5; 5
____ < 0 (true/false) | 6; false

____ point of Region C will satisfy the inequa- | No
lity. Only the points of Region ____ are in | B
the solution. The solution is written _____. | $-1 < x < 4$
Graph the solution.

4.7 Solving Quadratic Inequalities

3. $2x^2 + 5x - 3 < 0$

    Solve _____ = 0.

    ( )( ) = 0

    x = ____ or x = ____

    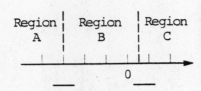

    $2x^2 + 5x - 3$

    $(2x - 1)(x + 3)$

    $\frac{1}{2}; -3$

    $-3; \frac{1}{2}$

    Try any point from Region A. The points of Region A (*do/do not*) satisfy the inequality. Try any point of Region B. The points of Region B (*do/do not*) satisfy the inequality. Try any point of Region C. The points of Region C (*do/do not*) satisfy the inequality. Write the solution.

    do not

    do

    do not

    _____

    Graph the solution.

    $-3 < x < \frac{1}{2}$

Solve each of the following inequalities and graph the solution.

4. $3m^2 - 7m - 6 > 0$ _____

    $m < -\frac{2}{3}$ or $m > 3$

    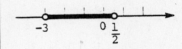

5. $6p^2 - 17p - 45 \leq 0$ _____

    $-\frac{5}{3} \leq p \leq \frac{9}{2}$

    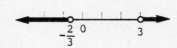

6. $(a + 2)(a - 3)(a + 1) \geq 0$ _____

    $-2 \leq a \leq -1$ or $a \geq 3$

    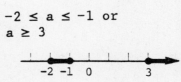

Chapter 4 Test

The answers for these questions are at the back of this Study Guide.

Factor completely.

1. $2xy + 16x^2y$

2. $12a^3b - 18ab^3$

3. $9r^2 - 6r^3 + 18r^4$

4. $32rs + 16r^2s + 48r^3$

5. $p^2 - 2p + 7p - 14$

6. $4 - 2m - 2q + qm$

7. $p^2 - 6p - 7$

8. $5p^4q^3 - 10p^3q^3 - 75p^2q^3$

9. $7m^2 + 22m + 3$

10. $10r^2 - 33r - 7$

11. $6x^2 + 7xy - 3y^2$

12. $21z^2 + 41zy + 10y^2$

13. $162p^2 - 50$

14. $25m^2 - 49q^2$

15. $z^4 - 81$

1. _____
2. _____
3. _____
4. _____
5. _____
6. _____
7. _____
8. _____
9. _____
10. _____
11. _____
12. _____
13. _____
14. _____
15. _____

16. $9p^2 - 30p + 25$

17. $64z^2 - 48z + 9$

18. $18z^3 - 24z^2 + 8z$

19. $27z^3 + 8$

20. $1000p^3 - 27q^6$

21. $x^2 - y^2 + 5x + 5y$

Solve each equation.

22. $6m^2 - 11m = 10$

23. $q(6q - 1) = 2$

24. $(3r + 2)(4r^2 + 7r - 15) = 0$

25. $r^3 - 25r = 0$

Solve each applied problem.

26. The length of a rectangle is 4 inches less than twice its width. The area is 96 square inches. Find the width of the rectangle.

27. The hypotenuse of a right triangle is 4 cm less than three times the shorter leg. The longer leg is 2 cm shorter than twice the smaller leg. Find the length of the shorter leg.

16. _____

17. _____

18. _____

19. _____

20. _____

21. _____

22. _____

23. _____

24. _____

25. _____

26. _____

27. _____

Chapter 4 Factoring and Applications

Graph the solution of each inequality.

28. $r^2 + 3r - 4 \leq 0$

28. ───────▶

29. $2k^2 + 5k - 3 > 0$

29. ───────▶

30. $10z^2 - 9z - 9 \geq 0$

30. ───────▶

5.1 The Fundamental Property of Rational Expressions

CHAPTER 5 RATIONAL EXPRESSIONS

5.1 The Fundamental Property of Rational Expressions

[1] Find the values for which a rational expression is undefined. (See Frames 1–6 below.)

[2] Find the numerical value of a rational expression. (Frames 7–10)

[3] Write rational expressions in lowest terms. (Frames 11–24)

---

1. A rational number is defined as the _____ of two _____, with the denominator not equal to ____. A rational expression is defined as the _____ of two _____, with denominator not equal to 0. For example,

   $$\frac{x^3}{x+1}, \quad \frac{x-6}{x+3}, \quad \frac{y^2-2y+4}{3y^2-4y+8}$$

   are all examples of _____ _____.

   | quotient
   | integers
   | 0
   | quotient;
   | polynomials
   |
   |
   | rational expressions

2. A rational expression is undefined for any value of the variable that makes the _____ equal to ___.

   | denominator
   | 0

3. For example, the rational expression

   $$\frac{9}{m-5}$$

   is undefined for m = ___, since this value makes the denominator equal ___.

   | 5
   | 0

4. To find the values for which a rational expression is undefined, use the following steps.

   1. Set the _____ of the rational expression equal to 0.

   2. _____ this equation.

   3. The solutions of the equation are the values which make the rational expression _____.

   | denominator
   |
   | Solve
   |
   | undefined

5. For what value of p is $\frac{1}{9p + 18}$ undefined?

   1. Set _____ equal to 0. | $9p + 18$
   2. Solve _____. | $9p + 18 = 0$
      $$9p = \_\_\_$$ | -18
      $$p = \_\_\_$$ | -2
   3. For p = ___, $\frac{1}{9p + 18}$ is undefined. | -2

6. To find the values for which
   $$\frac{8k + 1}{k^2 - 7k + 6}$$
   is undefined, solve the equation
   $$_____ = 0.$$ | $k^2 - 7k + 6$
   Factor: (   )(   ) = 0 | $k - 6$; $k - 1$

   The values that make the expression undefined are ___ and ___. | 6; 1

7. Rational expressions can be evaluated for given values of the variables. For example, if x = 4, then by substituting 4 for x, the rational expression
   $$\frac{3x^2 + 2}{4x + 1}$$
   becomes
   $$_____ = \_\_\_\_.$$ | $\frac{3(4)^2 + 2}{4(4) + 1}$; $\frac{50}{17}$

   If x = 3, then
   $$\frac{2x^2 - 4}{-3x + 9}$$
   becomes
   $$_____ = \_\_\_\_,$$ | $\frac{2(3)^2 - 4}{-3(3) + 9}$; $\frac{14}{0}$

   which has a denominator of ___. Thus, the rational expression (is/is not) defined when x = 3. | 0; is not

5.1 The Fundamental Property of Rational Expressions

Evaluate each expression in Frames 8–11 if $x = -2$.

8. $\dfrac{x-4}{x-6} = $ _____   Substitute -2 for x.

   $= $ _____   Subtract in numerator and denominator.

   $= $ _____   Lowest terms

   $\dfrac{-2-4}{-2-6}$

   $\dfrac{-6}{-8}$

   $\dfrac{3}{4}$

9. $\dfrac{9x^2 + 4x - 1}{2x^2 - 3x + 5} = $ _____

   $= $ _____

   $= $ _____

   $\dfrac{9(-2)^2 + 4(-2) - 1}{2(-2)^2 - 3(-2) + 5}$

   $\dfrac{36 - 8 - 1}{8 + 6 + 5}$

   $\dfrac{27}{19}$

10. $\dfrac{4x(x+3)}{2(x-3)} = $ _____

    $= $ _____

    $= $ _____

    $\dfrac{4(-2)(-2+3)}{2(-2-5)}$

    $\dfrac{-8}{-14}$

    $\dfrac{4}{7}$

11. $\dfrac{3x^2 - 4x + 6}{x^2 + 3x + 2} = $ ____, which (is/is not) defined.

    $\dfrac{26}{0}$; is not

12. A rational expression, just like a rational _____, can be written in _____ terms.   number; lowest

    We use the _____ _____ of rational expressions which says   fundamental property

    If P/Q is a rational expression and K represents any rational expression, where $K \neq 0$, then

    $\dfrac{PK}{QK} = $ _____.

    $\dfrac{P}{Q}$

    This property is based on the _____ property of multiplication.   identity

    $\dfrac{PK}{QK} = \dfrac{P}{Q} \cdot \dfrac{K}{K} = \dfrac{P}{Q} \cdot $ ___ $= $ ___

    1; $\dfrac{P}{Q}$

Chapter 5 Rational Expressions

13. Write $5x^2/10x$ in lowest terms, by factoring the numerator and denominator.

$$\frac{5x^2}{10x} = \underline{} \quad \textit{Factor.}$$

$\dfrac{5 \cdot x \cdot x}{5 \cdot 2 \cdot x}$

Then use the fundamental property of rational expressions.

$$= \underline{}$$

$\dfrac{x}{2}$

Write each rational expression of Frames 14–27 in lowest terms.

14. $\dfrac{5x(x-4)}{2(x-4)} = \underline{}$ $\qquad \dfrac{5x}{2}$

15. $\dfrac{8m^4}{12m^6} = \underline{}$ $\qquad \dfrac{2}{3m^2}$

16. $\dfrac{9p^5}{3p^7} = \underline{}$ $\qquad \dfrac{3}{p^2}$

17. $\dfrac{18x^2(x+2)}{12x^3(x+2)} = \underline{}$ $\qquad \dfrac{3}{2x}$

18. $\dfrac{6(m+4)}{8(m+4)^2} = \underline{}$ $\qquad \dfrac{3}{4(m+4)}$

19. To write $\dfrac{x^2 - 4x}{x - 4}$ in lowest terms, first _____ the numerator.   factor

$$\dfrac{x^2 - 4x}{x - 4} = \underline{} = \underline{}$$

$\dfrac{(x)(x-4)}{x-4}$; x

20. $\dfrac{r^3 - 9r^2}{r^2 - 9r} = \underline{} = \underline{}$ $\qquad \dfrac{(r^2)(r-9)}{r(r-9)}$; r

21. $\dfrac{x^2 - 16}{x + 4} = \underline{} = \underline{}$ $\qquad \dfrac{(x+4)(x-4)}{x+4}$; $x-4$

22. $\dfrac{m^2 - 25}{m - 5} = \underline{} = \underline{}$ $\qquad \dfrac{(m+5)(m-5)}{m-5}$; $m+5$

23. $\dfrac{k^2 - 5k + 6}{k - 3} = \underline{} = \underline{}$ $\qquad \dfrac{(k-2)(k-3)}{k-3}$; $k-2$

5.2 Multiplication and Division of Rational Expressions

24. $\dfrac{x^2 - x - 12}{x^2 - 7x + 12} =$ _____ = _____ | $\dfrac{(x-4)(x+3)}{(x-4)(x-3)}$; $\dfrac{x+3}{x-3}$

25. $\dfrac{a^2 + 5a - 6}{a^2 + a - 2} =$ _____ = _____ | $\dfrac{(a-1)(a+6)}{(a-1)(a+2)}$; $\dfrac{a+6}{a+2}$

26. $\dfrac{2z^2 - 7z - 4}{4z^2 - 17z + 4} =$ _____ = _____ | $\dfrac{(2z+1)(z-4)}{(4z-1)(z-4)}$; $\dfrac{2z+1}{4z-1}$

27. $\dfrac{r - s}{s - r} =$ _____ = _____ | $\dfrac{r-s}{-1(r-s)}$; -1

   Notice that ____ is a factor of the denominator. | -1
   The following rule may be stated.

   The quotient of two _____ expressions | nonzero
   that differ only in sign is ____. | -1

5.2 Multiplication and Division of Rational Expressions

[1] Multiply rational expressions. (See Frames 1-9 below.)

[2] Divide rational expressions. (Frames 10-22)

---

1. To multiply two rational expressions, _____ the numerators, and then _____ the denominators, and write the answer in _____ terms, if necessary. For example, | multiply

   multiply

   lowest

   $\dfrac{x^3}{4x^2} \cdot \dfrac{3x^5}{2x^7} =$ _____   *Multiply the numerators and the denominators.* | $\dfrac{x^3(3x^5)}{4x^2(2x^7)}$

   = _____ | $\dfrac{3x^8}{8x^9}$

   = _____   *Lowest terms* | $\dfrac{3}{8x}$

Multiply in Frames 2-9.

2. $\dfrac{(x-4)(x+3)}{(x+2)(x+3)} \cdot \dfrac{(x+2)(x-5)}{(x-4)(x+3)} =$ _____ | $\dfrac{x-5}{x+3}$

Chapter 5 Rational Expressions

3. $\dfrac{(2x-1)(x-5)}{(3x+2)(x+6)} \cdot \dfrac{(x+6)(x+3)}{(2x-1)(x-5)} =$ _____ $\dfrac{x+3}{3x+2}$

4. $\dfrac{a-2}{6} \cdot \dfrac{12}{a^2-4} =$ _____ Factor $\dfrac{a-2}{6} \cdot \dfrac{12}{(a+2)(a-2)}$

    = _____ $\dfrac{2}{a+2}$

5. $\dfrac{x^2-4}{x-2} \cdot \dfrac{x-2}{x+2} =$ _____ $\dfrac{(x+2)(x-2)}{x-2} \cdot \dfrac{x-2}{x+2}$

    = _____ $x-2$

6. $\dfrac{m^2-25}{m+5} \cdot \dfrac{m+5}{m-5} =$ _____ $\dfrac{(m+5)(m-5)}{m+5} \cdot \dfrac{m+5}{m-5}$

    = _____ $m+5$

7. $\dfrac{a^2+a-6}{a+3} \cdot \dfrac{a+2}{a^2-4} =$ _____ $\dfrac{(a+3)(a-2)}{a+3} \cdot \dfrac{a+2}{(a+2)(a-2)}$

    = _____ 1

8. $\dfrac{2m^2-m-3}{m^2-m-2} \cdot \dfrac{m-2}{m+1} =$ _____ $\dfrac{(2m-3)(m+1)}{(m+1)(m-2)} \cdot \dfrac{m-2}{m+1}$

    = _____ $\dfrac{2m-3}{m+1}$

9. $\dfrac{4k^2+8k+3}{2k^2-5k-12} \cdot \dfrac{3k^2-10k-8}{3k^2-16k-12}$

    = _____ $\dfrac{(2k+3)(2k+1)}{(2k+3)(k-4)} \cdot \dfrac{(k-4)(3k+2)}{(3k+2)(k-6)}$

    = _____ $\dfrac{2k+1}{k-6}$

10. To divide two rational expressions, _____ the   multiply
    first expressions by the _____ of the second   reciprocal
    expression. For example,

    $\dfrac{x^2-25}{x+4} \div \dfrac{x-5}{x-2}$ becomes $\dfrac{(x+5)(x-5)}{x+5} \cdot$ _____, $\dfrac{x-2}{x-5}$
    or, finally, _____. $x-2$

5.2 Multiplication and Division of Rational Expressions

Divide in Frames 11–18.

11. $\dfrac{(x-6)(2x+1)}{(x+3)(2x-5)} \div \dfrac{x-6}{2x-5}$

    $= \dfrac{(x-6)(2x+1)}{(x+3)(2x-5)} \cdot \underline{} = \underline{}$

    $\dfrac{2x-5}{x-6};\ \dfrac{2x+1}{x+3}$

12. $\dfrac{y-8}{y+2} \div \dfrac{y^2-6y+9}{y^2-y-6} = \dfrac{y-8}{y+2} \div \underline{}$

    $\dfrac{(y-3)(y-3)}{(y+2)(y-3)}$

    $= \dfrac{y-8}{y+2} \cdot \underline{}$

    $\dfrac{(y+2)(y-3)}{(y-3)(y-3)}$

    $= \underline{}$

    $\dfrac{y-8}{y-3}$

13. $\dfrac{x^2-3x+2}{x-2} \div \dfrac{x-4}{x^2-5x+4}$

    $= \underline{}$ *Factor*

    $\dfrac{(x-2)(x-1)}{x-2} \div \dfrac{x-4}{(x-4)(x-1)}$

    $= \underline{}$ *Multiply*

    $\dfrac{(x-2)(x-1)}{x-2} \cdot \dfrac{(x-4)(x-1)}{x-4}$

    $= \underline{}$

    $(x-1)^2$

14. $\dfrac{2m^2-m-3}{m+1} \div \dfrac{2m^2+5m-12}{m+3}$

    $= \underline{}$

    $\dfrac{(2m-3)(m+1)}{m+1} \cdot \dfrac{m+3}{(2m-3)(m+4)}$

    $= \underline{}$

    $\dfrac{m+3}{m+4}$

15. $\dfrac{3x^2-5x-2}{3x^2-10x+8} \div \dfrac{3x+1}{x+6}$

    $= \underline{}$

    $\dfrac{(3x+1)(x-2)}{(3x-4)(x-2)} \cdot \dfrac{x+6}{3x+1}$

    $= \underline{}$

    $\dfrac{x+6}{3x-4}$

168   Chapter 5   Rational Expressions

16. $\dfrac{m^2 - 3m - 10}{m^2 + 5m + 6} \div \dfrac{m^2 - 4m - 5}{m^2 - 3m - 18}$

   = _____

$\dfrac{(m-5)(m+2)}{(m+3)(m+2)} \cdot$

$\dfrac{(m+3)(m-6)}{(m+1)(m-5)}$

   = _____

$\dfrac{m-6}{m+1}$

17. $\dfrac{2k^2 - 5k - 3}{3k^2 - 10k + 3} \div \dfrac{2k^2 + 7k - 4}{3k^2 + 11k - 4} = $ _____

$\dfrac{2k+1}{2k-1}$

18. $\dfrac{6p^2 - p - 1}{8p^2 - 2p - 1} \div \dfrac{10p^2 + 11p - 6}{8p^2 + 14p + 3} = $ _____

$\dfrac{3p+1}{5p-2}$

19. Problems involving both multiplication and division can be worked by beginning at the _____, and doing all multiplications and divisions in order. For example, to work

$$\dfrac{x^2 - 2x}{x^2 + x - 6} \div \dfrac{x^2 - 4x}{x + 3} \div \dfrac{x}{x^2 - 5x + 4},$$

factor and find the reciprocals, working from the left:

left

_____

$\dfrac{x(x-2)}{(x+3)(x-2)} \cdot$

$\dfrac{x+3}{x(x-4)} \cdot$

$\dfrac{(x-4)(x-1)}{x}$

   = _____

$\dfrac{x-1}{x}$

Work the problems in Frames 20–22. Write all answers in lowest form.

20. $\dfrac{3r + 6}{16} \div \dfrac{9r + 18}{8} \cdot \dfrac{4}{r + 3}$

   = _____

$\dfrac{3(r+2)}{16} \cdot \dfrac{8}{9(r+2)} \cdot$

$\dfrac{4}{r+3}$

   = _____

$\dfrac{2}{3(r+3)}$

21. $\dfrac{2x-1}{3x+2} \cdot \dfrac{3x^2+2x}{8x^2+4x} \div \dfrac{2x-1}{2x+1} =$ _____ | $\dfrac{1}{4}$

22. $\dfrac{x^2-4x}{2x^2-7x-4} \cdot \dfrac{2x+1}{5x} \div \dfrac{2x^2}{15} =$ _____ | $\dfrac{3}{2x^2}$

5.3 The Least Common Denominator

[1] Find least common denominators. (See Frames 1–8 below.)

[2] Rewrite rational expressions with given denominators. (Frames 9–14)

1. To add or subtract fractions, we need a _____ common _____ .

   least
   denominator

2. A least common denominator is an expression that all the denominators of a problem will _____ into without a _____ . The abbreviation for least common denominator is ____ .

   divide
   remainder
   LCD

3. For example, the LCD for 3/4 and 2/5 is _____ , since ____ is the smallest number divisible by both ____ and ____ .

   20
   20
   4; 5

4. To find the LCD for 15m and 24m, write each in factored form.

   15m = _____
   24m = _____

   $3 \cdot 5 \cdot m$
   $2 \cdot 2 \cdot 2 \cdot 3 \cdot m$

   Take each factor the (*most/least*) times it appears.

   most

   LCD = ___ · ___ · ___ · ___ · ___ · ___
       = _____

   2; 2; 2; 3; 5; m
   120m

Chapter 5 Rational Expressions

Find the least common denominator for the rational expressions in Frames 5–8.

5. $\dfrac{1}{6y^2}, \dfrac{2}{27y^4}$

   Factor denominators. (Don't factor the
   _____.) | variables

   $6y^2 =$ ___ · ___ · y^2 | 2; 3
   $27y^4 =$ ___ · ___ · ___y^4 | 3; 3; 3

   Take the largest exponent on y, which is ___. | 4

   LCD = ___ · ___ · ___ · ___ · ___ | 2; 3; 3; 3; y^4
   = _____ | $54y^4$

6. $\dfrac{7}{10z}, \dfrac{3}{5z^2 - 15z}$

   Factor denominators.

   $10z =$ ___ · ___ · z | 2; 5
   $5z^2 - 15z = 5z($ $)$ | z − 3

   LCD = ___ · ___ · z · ( ) | 2; 5; z − 3
   = _____ | $10z(z-3)$

7. $\dfrac{8}{39m^5}, \dfrac{35m}{6m^3 - 9m^2}$

   $39m^5 =$ ___ · ___ · m^5 | 3; 13
   $6m^3 - 9m^2 =$ _____ | $3m^2(2m - 3)$
   LCD = _____ | $39m^5(2m - 3)$

8. $\dfrac{4}{m^2 - 5m + 4}, \dfrac{3}{m^2 - 6m + 8}$

   $m^2 - 5m + 4 = ($ $)($ $)$ | m − 4; m − 1
   $m^2 - 6m + 8 = ($ $)($ $)$ | m − 4; m − 2
   LCD = _____ | $(m-4)(m-1)(m-2)$

5.3 The Least Common Denominator   171

9. Once we find the least common denominator for a list of denominators, we can write each fraction with this _____.

   LCD, or least common denominator

10. The least common denominator for 1/4 and 5/6 is ___. Write each fraction with a denominator of 12.

    $$\frac{1}{4} = \frac{}{4 \cdot 3} = \frac{}{12}$$

    $$\frac{5}{6} = \frac{}{6 \cdot 2} = \frac{}{12}$$

    12

    $1 \cdot 3$; 3

    $5 \cdot 2$; 10

Rewrite each expression in Frames 11–14 with the given denominator.

11. $\dfrac{9}{8q} = \dfrac{}{16q^3}$

    Since $\dfrac{16q^3}{8q} = $ _____, multiply numerator and denominator by ____.

    $$\frac{9}{8q} = \frac{9 \cdot \_\_\_}{8q \cdot \_\_\_}$$

    $$= \frac{}{16q^3}$$

    $2q^2$

    $2q^2$

    $2q^2$
    $2q^2$
    $18q^2$

12. $\dfrac{8}{y + 7} = \dfrac{}{y^2 + 7y}$

    Since $\dfrac{y^2 + 7y}{y + 7} = \dfrac{y()}{y + 7} = $ ___,

    multiply numerator and denominator by ___.

    $$\frac{8}{y + 7} = \underline{}$$

    $$= \underline{}$$

    $y + 7$; y

    y

    $\dfrac{8 \cdot y}{(y + 7)y}$

    $\dfrac{8y}{y(y + 7)}$

13. $\dfrac{6r}{r^2 - 4r} = \dfrac{}{r(r - 4)(r + 1)}$

    $\dfrac{6r(r + 1)}{r(r - 4)(r + 1)}$

172   Chapter 5   Rational Expressions

14. $\dfrac{4p}{p^2 - 4p + 3} = \dfrac{}{(p - 3)(p - 1)(p + 7)}$	$\dfrac{4p(p + 7)}{(p - 3)(p - 1)(p + 7)}$

5.4   Addition and Subtraction of Rational Expressions

1   Add rational expressions having the same denominator. (See Frames 1–2 below.)

2   Add rational expressions having different denominators. (Frames 3–9)

3   Subtract rational expressions.   (Frames 10–16)

1 To add two rational expressions, they must both have the same _____. If both have the same denominator, then add the _____, and keep the same _____. For example,	denominator numerators denominator
$\dfrac{3}{k} + \dfrac{5}{k} = \dfrac{3 + 5}{k} = $ _____ .	$\dfrac{8}{k}$
2. Also,	
$\dfrac{m^2}{m + 2} + \dfrac{2m}{m + 2} = \dfrac{}{m + 2}$.	$m^2 + 2m$
Factor $m^2 + 2m$, and write the answer in lowest terms.	
$\dfrac{m^2 + 2m}{m + 2} = $ _____ *Factor* $\phantom{\dfrac{m^2 + 2m}{m + 2}} = $ _____ *Lowest terms*	$\dfrac{m(m + 2)}{m + 2}$ m
3. To add rational expressions with different denominators, find the _____. Then rewrite each rational expression as a fraction with the _____ as the denominator. Add the _____ of the two expressions. The _____ is the denominator of the sum.	LCD LCD numerators LCD

5.4 Addition and Subtraction of Rational Expressions

Find the sums in Frames 4–9. Write all answers in lowest terms.

4. $\dfrac{6}{r} + \dfrac{1}{2} = \dfrac{}{2r} + \dfrac{}{2r} = \dfrac{}{2r}$

    12; r; 12 + r

5. $\dfrac{2}{m+1} + \dfrac{3}{m} = \dfrac{}{m(m+1)} + \dfrac{}{m(m+1)}$

    2m; 3(m + 1)

    $= \dfrac{2m + 3()}{m(m+1)}$

    m + 1

    $= \dfrac{}{m(m+1)}$

    2m + 3m + 3

    $= \dfrac{}{m(m+1)}$

    5m + 3

6. $\dfrac{5}{x-2} + \dfrac{3}{x} = $ _____

    $\dfrac{8x - 6}{x(x - 2)}$

7. $\dfrac{4}{m-2} + \dfrac{3}{m+2} = $ _____

    $\dfrac{7m + 2}{(m-2)(m+2)}$

8. $\dfrac{3}{k^2 - 4} + \dfrac{2}{k+2} = \dfrac{3}{()()} + \dfrac{2}{k+2}$

    k + 2; k − 2

    $= \dfrac{3}{(k+2)(k-2)} + \dfrac{}{(k+2)(k-2)}$

    2(k − 2)

    $= \dfrac{}{(k+2)(k-2)}$

    2k − 1

9. $\dfrac{4}{x^2 + 2x - 3} + \dfrac{3}{x-1} = \dfrac{}{(x-1)(x+3)}$

    3x + 13

10. Subtraction is done in much the same way as addition. For example,

    $\dfrac{3}{k-1} - \dfrac{2(k+2)}{k-1} = \dfrac{3 - 2()}{k-1} = \dfrac{}{k-1}$.

    k + 2; −1 − 2k

Work the problems in Frames 11–16. Write each answer in lowest terms.

11. $\dfrac{8}{3k} - \dfrac{5}{4} = \dfrac{}{12k} - \dfrac{}{12k} = \dfrac{}{12k}$

    32; 15k; 32 − 15k

174   Chapter 5   Rational Expressions

12. $\dfrac{2}{m^2 - 16} - \dfrac{3}{m + 4} = \dfrac{2}{()()} - \dfrac{3}{m + 4}$ m − 4; m + 4

$= \dfrac{2 - 3()}{(m - 4)(m + 4)}$ m − 4

$= \dfrac{}{(m - 4)(m + 4)}$ 2 − 3m + 12

$= \dfrac{}{(m - 4)(m + 4)}$ −3m + 14

13. $\dfrac{5}{x - 1} - \dfrac{3}{x + 1} = \dfrac{}{(x - 1)(x + 1)}$ 2x + 8

14. $\dfrac{3x}{4x - 1} - \dfrac{2x}{4x + 1} = \dfrac{}{(4x - 1)(4x + 1)}$ 4x² + 5x

15. $\dfrac{4}{y + 3} - \dfrac{2}{(y + 3)(y - 1)} + \dfrac{1}{y - 1}$

$= \dfrac{4()}{(y + 3)(y - 1)} - \dfrac{2}{(y + 3)(y - 1)} + \dfrac{1()}{(y + 3)(y - 1)}$ y − 1; 6y + 3

$= \dfrac{}{(y + 3)(y - 1)}$ 5y − 3

16. $\dfrac{8}{p^2 - 7p + 6} - \dfrac{1}{p^2 - 3p - 18} = \underline{}$ $\dfrac{7p + 25}{(p - 6)(p - 1)(p + 3)}$

5.5 Complex Fractions

[1] Simplify a complex fraction by writing it as a division problem (Method 1). (See Frames 1-6 below.)

[2] Simplify a complex fraction by multiplying the numerator and the denominator by the LCD of all the fractions within the complex fraction (Method 2). (Frames 7-10)

---

1. A fraction containing fractions in its numerator, denominator, or both, is called a _____ fraction.   complex

5.5 Complex Fractions

2. To simplify the complex fraction

$$\frac{\frac{2}{3}+\frac{3}{4}}{\frac{5}{12}+\frac{1}{2}}$$

by Method 1, according to Step 1, write both numerator and denominator as single _____.   fractions

$$\frac{2}{3}+\frac{3}{4} = \frac{}{12}+\frac{}{12} = \frac{}{12}$$ 8; 9; 17

$$\frac{5}{12}+\frac{1}{2} = \frac{5}{12}+\frac{}{12} = \frac{}{12}$$ 6; 11

By Step 2, the original problem can now be written as a _____ problem.   division

$$\frac{\frac{2}{3}+\frac{3}{4}}{\frac{5}{12}+\frac{1}{2}} = \underline{}$$ $\frac{\frac{17}{12}}{\frac{11}{12}}$

Perform the indicated division, according to Step 3 by finding the _____ of the denominator and multiplying.   reciprocal

$$\frac{17}{12} \cdot \underline{} = \underline{}$$ $\frac{12}{11}$; $\frac{17}{11}$

Simplify each expression in Frames 3–4.

3. $\dfrac{\frac{3}{8}+\frac{5}{6}}{\frac{1}{2}-\frac{1}{3}}$

Here $3/8 + 5/6 = $ _____ and $1/2 - 1/3 = $ _____.   29/24; 1/6
Therefore,

$$\frac{\frac{3}{8}+\frac{5}{6}}{\frac{1}{2}-\frac{1}{3}} = \frac{\frac{29}{24}}{\frac{1}{6}} = \frac{29}{24} \cdot () = \underline{}.$$ $\frac{6}{1}$; $\frac{29}{4}$

Chapter 5 Rational Expressions

4. $\dfrac{\dfrac{12}{5} - \dfrac{5}{3}}{\dfrac{5}{6} + \dfrac{2}{5}} = $ _____

$\dfrac{22}{37}$

5. You can also simplify complex fractions involving variables. For example, to simplify

$$\dfrac{y + \dfrac{2}{y}}{y - \dfrac{2}{y}}$$

first write both numerator and denominator as single fractions.

$$y + \dfrac{2}{y} = \dfrac{y^2}{y} + \dfrac{2}{y} = \dfrac{}{y}$$

$y^2 + 2$

$$y - \dfrac{2}{y} = \dfrac{}{y}$$

$y^2 - 2$

Now change the complex fraction to a division problem and perform the indicated division.

$$\dfrac{y + \dfrac{2}{y}}{y - \dfrac{2}{y}} = \dfrac{\dfrac{y^2+2}{y}}{\dfrac{y^2-2}{y}} = \dfrac{y^2+2}{y} \cdot (\quad) = \underline{}$$

$\dfrac{y}{y^2-2}$; $\dfrac{y^2+2}{y^2-2}$

6. Simplify: $\dfrac{\dfrac{1}{m} + \dfrac{1}{n}}{\dfrac{1}{m} - \dfrac{1}{n}}$

Here $\dfrac{1}{m} + \dfrac{1}{n} = $ _____ and $\dfrac{1}{m} - \dfrac{1}{n} = $ _____.

$\dfrac{n+m}{mn}$; $\dfrac{n-m}{mn}$

Therefore, $\dfrac{\dfrac{1}{m} + \dfrac{1}{n}}{\dfrac{1}{m} - \dfrac{1}{n}} = $ _____.

$\dfrac{n+m}{n-m}$

7. By Method 2, complex fractions may also be simplified by multiplying both numerator and _____ by the least _____ denominator of all fractions in complex fraction.

denominator

common

8. For example, we can simplify

$$\frac{\frac{5}{9} + \frac{1}{2}}{\frac{5}{3} + \frac{3}{4}}$$

by multiplying both numerator and denominator by _____, since it is the LCD for 9, 2, _____, and _____. Do this.

36; 3
4

$$\frac{36\left(\frac{5}{9} + \frac{1}{2}\right)}{36(\quad)}$$

$\frac{5}{3} + \frac{3}{4}$

By the distributive property, this gives

$$\frac{36 \cdot \frac{5}{9} + \underline{}}{\underline{} + 36 \cdot \frac{3}{4}}$$

$36 \cdot \frac{1}{2}$

$36 \cdot \frac{5}{3}$

or _____ = _____.

$\frac{20 + 18}{60 + 27}$; $\frac{38}{87}$

Use Method 2 to simplify in Frames 9–10.

9. $\dfrac{k + \frac{1}{k}}{k - \frac{1}{k}} = $ _____

$\dfrac{k^2 + 1}{k^2 - 1}$

10. $\dfrac{\frac{1}{p^2} - \frac{1}{p}}{\frac{2}{p^2} + \frac{3}{p}} = $ _____

$\dfrac{1 - p}{2 + 3p}$

5.6 Equations Involving Rational Expressions

[1] Distinguish between expressions with rational coefficients and equations with terms that are rational expressions. (See Frames 1–9 below.)

[2] Solve equations with rational expressions. (Frames 10–18)

[3] Solve a formula for a specified variable. (Frames 19–22)

178   Chapter 5   Rational Expressions

1. $\frac{5}{4}x - \frac{2}{3}x$ is an _____ with rational | expression
   _____. It is the _____ | coefficients; difference
   of two terms. It can be _____ by finding | simplified
   the LCD, writing each coefficient with this LCD,
   and combining like terms. The LCD is _____. | 12
   Simplify the expression $\frac{5}{4}x - \frac{2}{3}x$.

   $\frac{5}{4}x - \frac{2}{3}x = (\quad)\frac{5}{4}x - (\quad)\frac{2}{3}x$ | $\frac{3}{3}; \frac{4}{4}$

   $= (\quad)x - (\quad)x$ | $\frac{15}{12}; \frac{8}{12}$

   $= (\quad)x.$ | $\frac{7}{12}$

2. $\frac{5}{4}x - \frac{2}{3}x = \frac{7}{4}$ is an _____. This is clear | equation
   because of the _____ sign. Solve by using | equal
   the _____ property of equality to | multiplication
   clear fractions. Multiply both sides by the
   _____, 12. | LCD

   $\frac{5}{4}x - \frac{2}{3}x = \frac{7}{4}$

   $\underline{\quad}(\frac{5}{4}x - \frac{2}{3}x) = \underline{\quad}(\frac{7}{4})$ | 12; 12

   $\underline{\quad}x - \underline{\quad}x = \underline{\quad}$ | 15; 8; 21

   $\underline{\quad}x = \underline{\quad}$ | 7; 21

   $x = \underline{\quad}$ | 3

3. When adding or subtracting, the LCD must be _____ | kept
   throughout the simplification. When solving an
   equation, the LCD is used to _____ both sides | multiply
   so that _____ are eliminated. | denominators

5.6 Equations Involving Rational Expressions 179

Identify an expression or an equation. If it is an expression, simplify it. If it is an equation, solve it.

4. $\frac{3}{5}x + \frac{1}{4}x = 17$

   This is an _____. It is _____ using the LCD, ____.   equation; solved 20

   $$\frac{3}{5}x + \frac{1}{4}x = 17$$

   $$(\quad)(\tfrac{3}{5}x + \tfrac{1}{4}x) = (\quad)17 \qquad 20;\ 20$$

   ___ x + ___ x = ____   12; 5; 340

   ___ x = ____   17; 340

   x = ____   20

5. $\frac{9}{4}y + \frac{1}{6}y$

   This is an _____. It is _____ using the LCD, _____.   expression; simplified 12

   $$(\quad)\tfrac{9}{4}y + (\quad)\tfrac{1}{6}y \qquad \tfrac{3}{3};\ \tfrac{2}{2}$$

   $$= (\quad)y + (\quad)y \qquad \tfrac{27}{12};\ \tfrac{2}{12}$$

   $$= (\quad)y \qquad \tfrac{29}{12}$$

6. $\frac{1}{3}y + \frac{3}{5}y = \frac{14}{3}$

   This is an _____.   equation

   y = ____   5

7. $\frac{7}{8}z - \frac{2}{3}z$

   This is an _____.   expression

   $\frac{7}{8}z - \frac{2}{3}x = $ _____ $\frac{5}{24}z$

180 Chapter 5 Rational Expressions

8. $\frac{9}{4}x + \frac{1}{2}x$

   This is an _____.   expression

   $\frac{9}{4}x + \frac{1}{2}x =$ _____ $\frac{11}{4}x$

9. $\frac{5}{6}z - \frac{1}{5}z = \frac{19}{2}$

   This is an _____.   equation

   $z =$ ____.   15

Solve the equations in Frames 10–18. Be sure to check your answers.

10. $\frac{4}{x} = \frac{3}{8}$

    Multiply both sides by ____. $8x$

    $8x\left(\frac{4}{x}\right) = 8x(\quad)$ $\frac{3}{8}$

    ____ = ____   32; 3x

    $x =$ ____ $\frac{32}{3}$

11. $\frac{3}{x} - \frac{5}{x} = -2$

    Multiply both sides by ____. x

    $x\left(\frac{3}{x} - \frac{5}{x}\right) =$ ____ $-2x$

    By the _____ property,   distributive

    $x\left(\frac{3}{x}\right) - x(\quad) =$ ____ $\frac{5}{x}$; $-2x$

    ____ − ____ = $-2x$ 3; 5

    ____ = $-2x$ -2

    $x =$ ____.   1

12. $\frac{2}{m-4} = \frac{10}{m}$

    Multiply both sides by _____. $m(m-4)$

    $m =$ ____   5

5.6 Equations Involving Rational Expressions

13. $\dfrac{k+4}{6} = \dfrac{k-4}{2}$ k = ____ | 8

14. $\dfrac{m}{m+6} = \dfrac{-6}{m+6} - 1$ m = ____ | No solution (Did you check your possible solution?)

15. $\dfrac{3}{2k-2} + \dfrac{4}{3k-3} = \dfrac{17}{6}$

    Factor all denominators.

    $\dfrac{3}{\rule{1cm}{0.15mm}} + \dfrac{4}{\rule{1cm}{0.15mm}} = \dfrac{17}{6}$ | $2(k-1)$; $3(k-1)$

    Multiply both sides by ____, obtaining | $6(k-1)$

    ____ + ____ = ____ | 9; 8; $17(k-1)$

    $17 = 17k -$ ____ | 17

    ____ $= 17k$ | 34

    $k =$ ____. | 2

16. $\dfrac{2}{p-5} - \dfrac{3}{2p-10} = -\dfrac{1}{8}$ p = ____ | 1

17. $\dfrac{3}{m-1} + \dfrac{4}{m+1} = \dfrac{13}{(m-1)(m+1)}$ m = ____ | 2

18. $\dfrac{5}{x^2+x-2} - \dfrac{10}{x^2+2x-3} = \dfrac{-3}{x^2+5x+6}$

    Factor all denominators.

    $\dfrac{5}{(\rule{0.5cm}{0.15mm})(\rule{0.5cm}{0.15mm})} - \dfrac{10}{(x-1)(\rule{0.5cm}{0.15mm})}$ | $(x-1)(x+2)$; $(x+3)$

    $= \dfrac{-3}{(x+2)(x+3)}$

    Multiply through by ____. | $(x-1)(x+2)(x+3)$

    $5(\rule{0.5cm}{0.15mm}) - 10(\rule{0.5cm}{0.15mm}) = -3(\rule{0.5cm}{0.15mm})$ | $x+3$; $x+2$; $x-1$

    x = ____ | −4

19. Sometimes a formula is given and we must solve the formula for one of its ____. This process is called ____ for a ____ variable. | variables

    solving; specified

182   Chapter 5 Rational Expressions

20. For example, to solve

$$\frac{2}{p} - \frac{5}{q} = \frac{1}{4}$$

for q, multiply both sides of the equation by

_____.   4pq

$$4pq\left(\frac{2}{p} - \frac{5}{q}\right) = \underline{}$$ $4pq\left(\frac{1}{4}\right)$

$$4pq\left(\frac{2}{p}\right) - 4pq() = \underline{}$$ $\frac{5}{q}$; pq

$$8q - \underline{} = pq$$ 20p

Since we solving for ____, collect all terms with   q
q on one side of the equation.

$$8q - \underline{} = \underline{}$$ pq; 20p

On the left, factor out _____.   q

$$\underline{} = 20p$$ q(8 - p)

Divide both sides by _____, to get   8 - p

$$q = \underline{}.$$ $\frac{20p}{8 - p}$

In Frames 21–22, solve for x.

21. $\frac{6}{a} + \frac{3}{x} = 1$ x = _____ $\frac{3a}{a - 6}$

22. $m = \frac{2 + p}{1 + x}$ x = _____ $\frac{2 + p - m}{m}$

5.7 Applications of Rational Expressions

[1] Solve applications with rational expressions involving unknown numbers. (See Frames 1–3 below.)

[2] Solve applications with rational expressions involving distance. (Frames 4–5)

[3] Solve applications with rational expressions involving work. (Frames 6–8)

[4] Solve problems about variation. (Frames 9–12)

5.7 Applications of Rational Expressions

1. If the same number is added to the numerator and denominator of the fraction 5/8, the result is 1/2. Find the number that must be added.

   To identify the variable, look for a phrase such as "how many," "how far," or "find." Here we have the phrase "find." Thus, x = _____ _____. (We could use any letter.) The statement "the same number is added to the numerator" is written in symbols as _____; and "the same number is added to the denominator" is written _____. Since "the result is _____," then

   $$_____ = \_\_\_\_.$$

   Multiplying both sides by the LCD, _____, gives

   $$_____ = _____$$
   $$10 + \_\_\_ = 8 + x$$
   $$x = \_\_\_\_.$$

   To a check, substitute -2 for x in the original problem.

   $$_____ = \_\_\_\_ \quad (true/false)$$

the number that must be added	
$5 + x$	
$8 + x$; $1/2$	
$\dfrac{5+x}{8+x}$; $\dfrac{1}{2}$	
$2(8 + x)$	
$2(5 + x)$; $8 + x$	
$2x$	
-2	
$\dfrac{5+(-2)}{8+(-2)}$; $\dfrac{1}{2}$; true	

2. If the same number is added to both the numerator and denominator of the fraction 4/9, the result is 2/3. Find the number.

   Here x = _____.   the number to be added

   The equation expressing the problem in symbols is

   $$_____ = \dfrac{2}{3}.$$ $\dfrac{4+x}{9+x}$

   Solve this equation.

   $$x = \_\_\_\_$$ 6

184   Chapter 5   Rational Expressions

3. If twice a number is added to the reciprocal of 2, the result is 7/2. Find the number.

 Let x = _____. | the number desired

  "Twice a number" is written ____. | 2x

  "The reciprocal of 2" is written ____. | 1/2

 Write the equation.

$$_____ = \frac{7}{2}$$ | $2x + \frac{1}{2}$

 Solve the equation.

$$x = \_\_\_\_$$ | $\frac{3}{2}$

4. A boat goes 10 miles per hour in still water. It can go 24 miles upstream in the same time as 36 miles downstream. Find the speed of the current.

 Let x = _____. The basic formula for distance problems is | the speed of the current

  distance = _____ · _____, | rate; time
  or d = ____. | rt

If the boat goes 10 miles per hour in still water, and the speed of the current is ____ miles per hour, then the speed upstream is _____ and the speed downstream is _____. Write the distance upstream, distance downstream, speed upstream, and speed downstream in the chart that follows. | x
 | 10 − x
 | 10 + x

To find the time, use the formula d = rt. Divide both sides of the formula by ____. Then t = ____. Therefore, the time upstream is | r; d/r

$$t = \frac{d}{r} = _____$$ | $\frac{24}{10-x}$

and the time downstream is _____. Write these in the chart that follows. | $\frac{36}{10+x}$

5.7 Applications of Rational Expressions

	d	r	t
Upstream			
Downstream			

d	r	t
24	10 − x	$\frac{24}{10-x}$
36	10 + x	$\frac{36}{10+x}$

The problem says that the times are equal, so that

_____ = _____ .

$\frac{24}{10-x}$; $\frac{36}{10+x}$

Solve for x.

x = ____

2

The speed of the current is ___ miles per hour.

2

5. Joe Swell flew to the Mustang Recreation Area at 125 miles per hour. The return trip, because of head winds, took one hour longer and was at 100 miles per hour. Find the distance to Mustang.

Let x = _____ . Complete the following chart. (Don't forget: t = ____ .)

the distance
d/r

	d	r	t
To Mustang			
From Mustang			

d	r	t
x	125	$\frac{x}{125}$
x	100	$\frac{x}{100}$

Since the return trip took one hour longer,

$$\frac{x}{100} = \frac{x}{125} + \underline{}$$

1

Multiply both sides by 500 and solve for x.

x = ____

500

The distance to Mustang is _____ .

500 miles

Chapter 5 Rational Expressions

6. **Samantha can take inventory at her card shop in 7 hours when she works alone, while Harriet can do the same job in 9 hours. How long will it take them to do this job when working together?**

   Problems of this type can be worked by considering the fractional part of the work done by each worker.

   Let x = _____.   the number of hours it will take them together to take inventory

   Samantha's rate alone is _____ of the job per hour and Harriet's rate is _____ of the job per hour. Make a chart.   1/7   1/9

Worker	Rate	Time working together	Fractional part done when working together
Samantha			
Harriet			

Rate	Time	Fractional part
$\frac{1}{7}$	x	$\frac{1}{7}x$
$\frac{1}{9}$	x	$\frac{1}{9}x$

   Since they do 1 whole job working together,

   fractional part done by Samantha + fractional part done by Harriet = 1 whole job.

   ↓   ↓   ↓

   _____ + _____ = _____ $\frac{1}{7}x$; $\frac{1}{9}x$; 1

   Multiply by the LCD, ____, to solve this equation.   63

                                        x = ____ $\frac{63}{16}$

   (63/16 is close to ____ hours.)   4

7. **The Youth Group can clean the trash from a park in 5 hours, while the Young Adults need 6 hours. Find the time it would take for both groups, working together.**

5.7 Applications of Rational Expressions 187

Let x = _____.   the number of hours
                                  it would take both
                                  groups working to-
                                  gether

The chart follows.

Group	Rate	Time working together	Fractional part done when working together
Youth Group			
Young Adults			

Rate	Time	Fractional part
$\frac{1}{5}$	x	$\frac{1}{5}x$
$\frac{1}{6}$	x	$\frac{1}{6}x$

Solve the equation

_____ + _____ = _____ . $\frac{1}{5}x$; $\frac{1}{6}x$; 1

                               x = ____ $\frac{30}{11}$

It would take _____ for both groups   30/11 hours
working together.

8. **An inlet pipe can fill a pickle barrel in 9 hours, while a drain can empty it in 15 hours. Find the time it would take to fill the barrel if both the inlet and drain are open.**

Let x = _____.   the number of hours
                                  to fill the barrel
                                  with both inlet and
                                  drain open

The rate that the inlet pipe would fill the barrel
per hour is _____. The rate that the drain would   1/9
empty the barrel per hour is _____.   1/15
Make a chart.

	Rate	Time working together	Fractional part done when working together
Inlet			
Drain			

Rate	Time	Fractional part
$\frac{1}{9}$	x	$\frac{1}{9}x$
$\frac{1}{15}$	x	$\frac{1}{15}x$

Chapter 5 Rational Expressions

The equation is

$$\underline{\hspace{2em}} - \underline{\hspace{2em}} = \underline{\hspace{2em}}.$$ $\frac{1}{9}x$; $\frac{1}{15}x$; 1

A minus sign is used because the inlet and drain are working _____ each other.   against

Solve the equation,

$$x = \underline{\hspace{2em}}$$ $\frac{45}{2}$

It would take _____ with both pipes open.   45/2 hours

9. Variation shows how one variable changes as another does. Fill in the following blanks where k is a constant.

   y varies _____ as x if $y = kx$.   directly
   y varies inversely as x if _____ $y = k/x$
   y varies as the cube of x if _____ $y = kx^3$

10. Suppose y varies inversely as x, and $y = 6$ when $x = 2$. Find y when $x = 9$.

    Since y varies inversely as x, there is a constant k such that

    $$\underline{\hspace{6em}}.$$ $y = \frac{k}{x}$

    To find the value of k, substitute ____ for y and 2 for ___.   6
    x

    $$6 = \underline{\hspace{2em}}$$ $\frac{k}{2}$

    Solve for k:

    $$k = \underline{\hspace{2em}}$$ 12

    Find y when $x = 9$.

    $$y = \underline{\hspace{2em}}$$ $\frac{12}{9} = \frac{4}{3}$

5.7 Applications of Rational Expressions

Solve each variation problem in Frames 11–12.

11. Suppose m varies directly as p^2. Suppose m = 96 when p = 4. Find m when p is 15.

    _____   1350

12. Suppose r varies inversely as the square of t, and r = 48 when t = 1/2. Find r when t = 2/3.

    $$r = \underline{\hspace{1cm}}$$ $\dfrac{k}{t^2}$

    Let r = ____ and t = ____.   48; 1/2

    $$48 = \frac{k}{\left(\frac{1}{2}\right)^2}$$

    $$48 = \frac{k}{\frac{1}{4}}$$

    $$48 = \underline{\hspace{1cm}}$$ 4k

    $$\underline{\hspace{1cm}} = k$$ 12

    Now,

    $$r = \underline{\hspace{2cm}}.$$ $\dfrac{12}{t^2}$

    Replace t with ____ to get r = ____.   2/3; 27

Chapter 5 Test

The answers for these questions are at the back of this Study Guide.

1. Find any values for which $\dfrac{3r - 5}{r^2 - 2r - 15}$ is undefined.

   1. _____

2. Find the numerical value of

   $$\dfrac{2z - 1}{z^2 - z - 4}$$

   when (a) $z = -1$ and (b) $z = 2$.

   2. (a) _____

   (b) _____

Write each rational expression in lowest terms.

3. $\dfrac{7q^4 r^7}{14 q^6 r}$

4. $\dfrac{3r^2 + r - 2}{6r - 4}$

   3. _____

   4. _____

Multiply or divide, as indicated. Write all answers in lowest terms.

5. $\dfrac{p^5 q}{p^2} \cdot \dfrac{p^3}{p^7 q^2}$

   5. _____

6. $\dfrac{3a - 15}{4} \div \dfrac{10 - 2a}{6}$

   6. _____

7. $\dfrac{r^2 - r - 6}{r^2 - 8r + 15} \cdot \dfrac{r^2 + 2r - 35}{r^2 + 13r + 42}$

   7. _____

8. $\dfrac{2p^2 + 3p - 2}{3p^2 + 7p + 2} \div \dfrac{4p^2 - 12p + 5}{6p^2 + 17p + 5}$

   8. _____

Find the least common denominator for the following fractions.

9. $\dfrac{5}{8y^2},\ \dfrac{1}{12y^3},\ \dfrac{-2}{9y}$

   9. _____

10. $\dfrac{m + 1}{2m^2 + m - 6},\ \dfrac{2m - 3}{6m^2 - 17m + 12}$

    10. _____

Rewrite each rational expression with the given denominator.

11. $\dfrac{8}{3z} = \dfrac{}{12z^4}$

11. _____

12. $\dfrac{2}{5m - 10} = \dfrac{}{15m^2 - 30m}$

12. _____

Add or subtract as indicated. Write all answers in lowest terms.

13. $\dfrac{2}{q} - \dfrac{5}{q}$

13. _____

14. $\dfrac{-2}{m + 3} + \dfrac{4}{5m + 15}$

14. _____

15. $\dfrac{a}{a - 1} + \dfrac{a + 2}{a - 2}$

15. _____

16. $\dfrac{2}{2m^2 + 3m - 2} - \dfrac{m}{2m^2 + m - 6}$

16. _____

Simplify each complex fraction.

17. $\dfrac{\dfrac{1}{r} + 5}{\dfrac{1}{r} + 3}$

18. $\dfrac{\dfrac{1}{r - 1} + 1}{\dfrac{2}{r - 1} - 3}$

17. _____

18. _____

Solve each equation.

19. $\dfrac{4}{7z} + \dfrac{2}{3z} = \dfrac{26}{21}$

19. _____

20. $\dfrac{y}{y + 1} = 2 - \dfrac{1}{y + 1}$

20. _____

21. $\dfrac{3}{y^2 - y - 2} = \dfrac{2}{y + 1} + \dfrac{3}{y - 2}$

21. _____

Chapter 5 Rational Expressions

Solve each problem.

22. If three times a number is added to twice the reciprocal of the number, the result is 5. Find the number.

22. _____

23. The current in a river is 4 miles per hour. A boat can go 56 miles downstream in the same time as 24 miles upstream. Find the speed of the boat in still water.

23. _____

24. A man can paint a room in his house, working alone, in 8 hours. His wife can do the job in 5 hours. How long will it take them to paint the house if they work together?

24. _____

25. If x varies directly as y, and $x = 14$ when $y = 2$, find x when $y = 7$.

25. _____

CHAPTER 6 GRAPHING LINEAR EQUATIONS

6.1 Linear Equations in Two Variables

1. Write a solution as an ordered pair. (See Frames 1–4 below.)
2. Decide whether a given ordered pair is a solution of a given equation. (Frames 5–8)
3. Complete ordered pairs for a given equation. (Frames 9–12)
4. Complete a table of values. (Frames 13–15)
5. Plot ordered pairs. (Frames 16–22)

---

1. You solved equations in Chapter 2 which have only one variable, equations such as $2x + 4 = 12$ and $5x - 3 = 8$. However, many practical problems involve equations in two variables, for example, $x + 2y = 11$. The equations of this section can all be written in the form. $Ax + By = C$, where A, B, and C are real numbers and A and B are not both 0. Such equations are called _____ equations in two variables.

	linear

    Is $4x + 5y = 9$ a linear equation? (*yes/no*) — yes
    Is $x^2 = 36$ a linear equation? (*yes/no*) — no
    Is $x = 6$ a linear equation? (*yes/no*) — yes

2. One solution for the equation $x + 2y = 11$ is $x = 3$ and $y = 4$. To see this, substitute ____ for x and ____ for y in the equation $x + 2y = 11$.

    ____ + 2(____) = 11

    _____ = 11 (*true/false*)

    Answers: 3; 4; 3; 4; 11; true

3. It is a lot of trouble to write the solution you found in Frame 2, as $x = 3$ and $y = 4$. A shorthand way is to write this solution as

    (,),   (3, 4)

Chapter 6 Graphing Linear Equations

where the ____ value is given first. The symbol (3, 4) is an example of an _____ pair of numbers.

x	
ordered	

4. If you choose a value of x, say x = −5, you can find the corresponding value of y for the equation x + 2y = 11. To do so, substitute ____ for x, and solve for ____.

$$\underline{} + 2y = 11$$
$$2y = \underline{}$$
$$y = \underline{}$$

The ordered pair (,) is a solution for x + 2y = 11. Complete the following ordered pairs for the equation x + 2y = 11.

(−9,)
(−3,)
(, 0)
(0,)

These examples suggest that x + 2y = 11 is satisfied by a(n) (*infinite/finite*) number of ordered pairs.

−5
y
−5
16
8
(−5, 8)
10
7
11
11/2
infinite

5. You found in Frame 3 above that the ordered pair (3, 4) is a solution of the equation x + 2y = 11. However, there are many other pairs of numbers which are also solutions. For example, to check that (7, 2) is a solution, replace x with ____ and y with ____ in the equation x + 2y = 11.

$$x + 2y = 11$$
$$\underline{} + 2(\underline{}) = 11$$
$$\underline{} = 11 \quad (\textit{true/false})$$

Therefore, (7, 2) (*is/is not*) a solution of x + 2y = 11.

7
2
7; 2
11; true
is

6.1 Linear Equations in Two Variables

6. Is (-3, 2) a solution for x + 2y = 11? To find out, replace x with ____ and y with ____. | -3; 2

   ____ + 2( ) = 11 | -3; 2

   ____ = 11 (true/false) | 1; false

   (-3, 2) (is/is not) a solution of x + 2y = 11. | is not

7. Is (-5, 8) a solution for 4x + y = -28?

   (yes/no) | no

8. Is (7, 2) a solution for x = 7? (yes/no) | yes

9. Let 3x - 2y = 6, and complete the following ordered pairs.

   (0,)   Let ____ = 0. Then | x

   3( ) - 2y = 6 | 0

   y = ____. | -3

   The ordered pair is (0,   ). | -3

   (, 0) | 2

   (-2,) | -6

   (, 3) | 4

   (, 9) | 8

In Frames 10–12, complete the ordered pairs.

10. x + y = 12

x	y	Ordered pair	
8	____	(,)	4; (8, 4)
0	____	(,)	12; (0, 12)
____	0	(,)	12; (12, 0)
____	-2	(,)	14; (14, -2)
____	9	(,)	3; (3, 9)

Chapter 6 Graphing Linear Equations

11. $4x - 3y = 12$

x	y	Ordered pair	
0	___	(,)	−4; (0, −4)
___	0	(,)	3; (3, 0)
−3	___	(,)	−8; (−3, −8)
___	4	(,)	6; (6, 4)
___	2	(,)	9/2; (9/2, 2)
2	___	(,)	−4/3; (2, −4/3)

12. $x = 5$

x	y	Ordered pair	
___	2	(,)	5; (5, 2)
___	0	(,)	5; (5, 0)
___	3	(,)	5; (5, 3)
___	−5	(,)	5; (5, −5)
___	−12	(,)	5; (5, −12)

13. Ordered pairs of an equation are often displayed in a table of values. The table may be written vertically or _____, as in the next frame.   horizontally

14. Complete the table of values for the equation $2x + y = 10$.

x	1	6		
y			0	−4

Complete the first ordered pair by letting x = ___.

$2x + y = 10$
$2(\ ) + y = 10$
___ $+ y = 10$
$y = 10$ ___
$y = $ ___

1

1
2
−2
8

Then, let x = ___ . | 6

  2( ) + y = 10 | 6

  ___ + y = 10 | 12

         y = ___ | −2

Complete the third ordered pair by letting
___ = 0. | y

  2x + ___ = 10 | 0

  2x = ___ | 10

   x = ___ | 5

Then, let y = ___. | −4

  2x + ( ) = 10 | −4

  2x = 10 + ___ | 4

  2x = ___ | 14

   x = ___ . | 7

The completed table of values is as follows.

x	1	6		
y			0	−4

                                                    | 5; 7
                                                    | 8; −2

15. Complete the table of values for the equation
    y − 4 = 0.

x	−1	0	4	6
y				

                                                    | 4; 4; 4; 4

If you think of the equation y − 4 = 0 as an
equation in two variables, it might be written

  ___ · x + y = ___ . | 0; 4

This form of the equation shows that for any
value of ___, the value of y is ___. | x; 4

16. To graph ordered pairs, we use a _____ | coordinate
    system such as the one given at the top of the
    next page. The horizontal line is called the
    the _____, and the vertical line is called | x-axis
    the _____. The point where the horizontal | y-axis
    and vertical lines meet is called the _____. | origin

198 Chapter 6 Graphing Linear Equations

To graph the ordered pair (2, 5), go ____ units to the _____ on the x-axis, and then go ____ units up and parallel to the y-axis.

2

right; 5

Graph the point (2, 5) on the coordinate system.

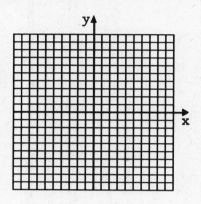

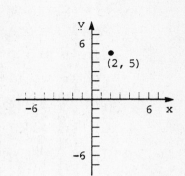

17. Graph each of the following ten points on the coordinate system above. Answers are at the right.

(3, -2)
(5, 2)
(-3, 2)
(-2, -3)
(-3, -4)
(2, -4)
(-1, 0)
(-5, 0)
(0, 3)
(0, -2)

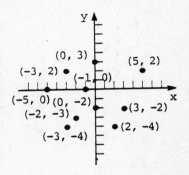

18. Identify each of the lettered points on the coordinate system below.

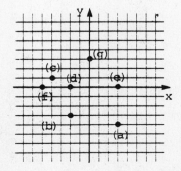

(a) _____   (3, -4)
(b) _____   (-2, -3)
(c) _____   (-4, 1)
(d) _____   (-2, 0)
(e) _____   (3, 0)
(f) _____   (-5, 0)
(g) _____   (0, 3)

6.1 Linear Equations in Two Variables 199

19. Identify each of the four quadrants on the coordinate system at the right. The points on the x-axis and the y-axis belong to _____ quadrant.

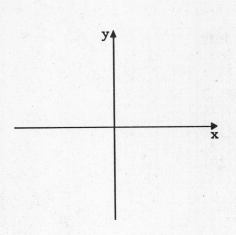

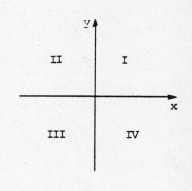

no

20. Complete the table of values for the equation x + y = 8.

x	0	2	4	6	8
y					

    Graph these ordered pairs.

    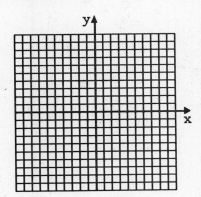

    8; 6; 4; 2; 0

    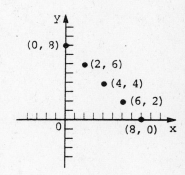

21. Complete the table of values for the equation 2x − 3y = 12, and then graph the ordered pairs.

x	−3	0	3	6
y				

    −6; −4; −2; 0

200 Chapter 6 Graphing Linear Equations

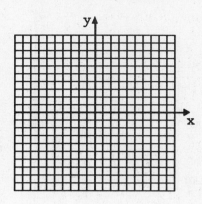

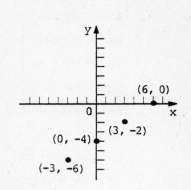

22. Complete the table of values for the equation $4y - x = 9$ and then graph the ordered pairs.

x	-1	0	3	7
y				

2; 9/4; 3; 4

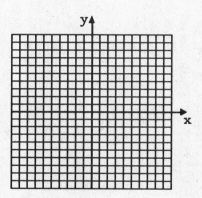

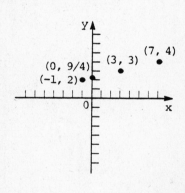

6.2 Graphing Linear Equations in Two Variables

1. Graph linear equations by completing and plotting ordered pairs. (See Frames 1-3 below.)

2. Find intercepts. (Frames 4-7)

3. Graph linear equations of the form $Ax + By = 0$. (Frames 8-10)

4. Graph linear equations of the form $y = k$ or $x = k$. (Frames 11-12)

6.2 Graphing Linear Equations in Two Variables

1. A vertical _____ of values can be used to help plot points for a _____.   **table**
   graph

   Complete the given table of values for the equation $x + 2y = 11$ and then plot these points on the graph.

   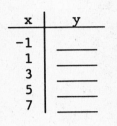
   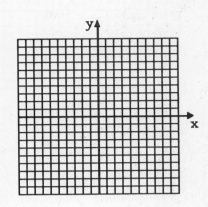
   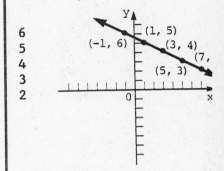

   The points on your graph should all lie on the same straight line. Draw a straight line through your points.

2. To find the straight line which is the graph of $3x - 2y = 12$, it is probably a good idea to locate at least _____ points of the graph. To find each of these three points, choose a value for x (or y). Then find the corresponding value of the other variable. Complete the table of values below. Then plot the points and draw a straight line through them.   **three**

   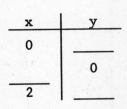
   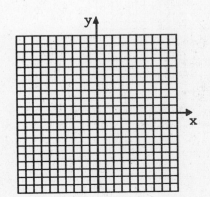
   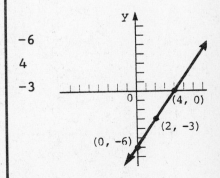

   −6
   4
   −3

202 Chapter 6 Graphing Linear Equations

3. Graph $2x - 5y = 10$.

x	y
	0
0	
-5	

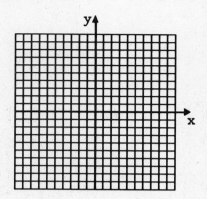

5
-2
-4

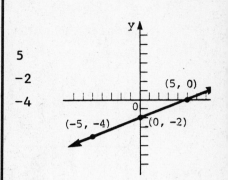

4. One good way to graph many lines is to find the intercepts. The x-intercept of a line is the point where the line crosses the _____, while the y-intercept is the point where the line crosses the _____. To find the x-intercept, let _____ and solve for x. To find the y-intercept, let _____ and solve for y. Find the intercepts for $x + y = -4$.

x-axis

y-axis

$y = 0$

$x = 0$

   x-intercept: _____   (-4, 0)
   y-intercept: _____   (0, -4)

Use the point (-2,   ) as a check.   -2
Complete the graph of $x + y = -4$.

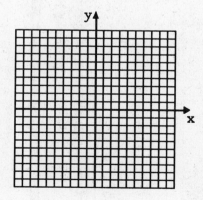

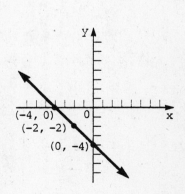

6.2 Graphing Linear Equations in Two Variables   203

Use intercepts to graph the lines in Frames 5—7.

5. $3x - 7y = 21$

    x-intercept: _____   (7, 0)
    y-intercept: _____   (0, -3)
    Check point: (3,)   -12/7

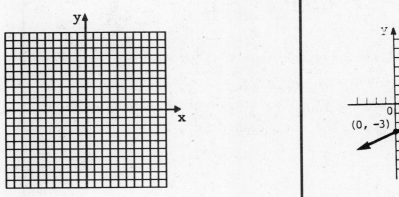

6. $3x + 2y = 6$

    x-intercept: _____   (2, 0)
    y-intercept: _____   (0, 3)
    Check point: (4,)   -3

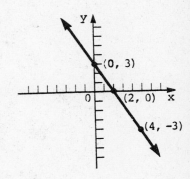

7. $x - 2y = -1$

    x-intercept: _____   (-1, 0)
    y-intercept: _____   (0, 1/2)
    Check point: (5,)   3

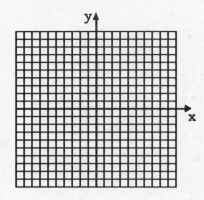

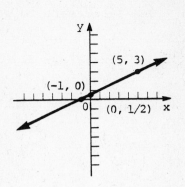

8. Some lines have the same x-intercept and y-intercept: the point _____. In this case, it is necessary to find two other points on the line. Do this to find the graph of y = 3x.

x	y
0	__
2	__
-2	__

(0, 0)

0
6
-6

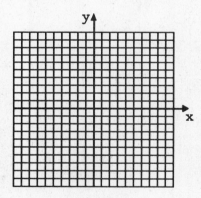

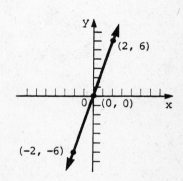

Graph each of the following.

9. y = -2x

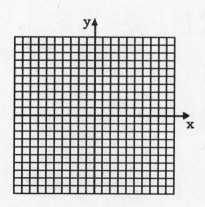

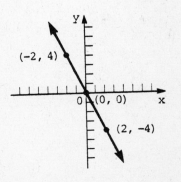

6.2 Graphing Linear Equations in Two Variables   205

10. $x - 4y = 0$

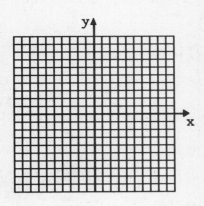

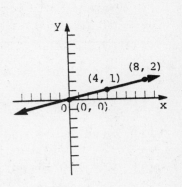

11. To graph $y = 4$, complete the table of values below. Then draw the graph.

x	y
2	___
0	___
-3	___

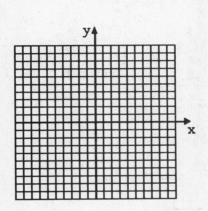

4
4
4

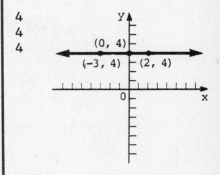

The graph is a (*horizontal/vertical*) straight line.

horizontal

12. To graph $x = 3$, complete the table of values below. Then draw the graph.

x	y
___	2
___	0
___	-4

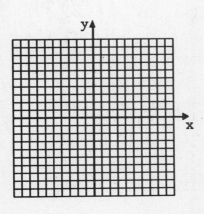

3
3
3

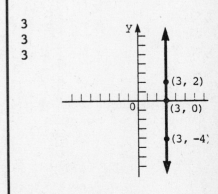

Chapter 6 Graphing Linear Equations

The graph is a (*vertical/horizontal*) straight line.	vertical

6.3 The Slope of a Line

[1] Find the slope of a line given two points. (See Frames 1-7 below.)

[2] Find the slope from the equation of a line. (Frames 8-12)

[3] Use the slope to determine whether two lines are parallel, perpendicular, or neither. (Frames 13-16)

---

1. Slope is used to measure the _____ of a line. — steepness

2. Slope is defined as

   $$\text{slope} = \frac{\text{change in \_\_\_\_\_ values}}{\text{change in \_\_\_\_\_ values}}.$$

   y
   x

3. If a line goes through the points (x_1, y_1) and (x_2, y_2), its slope is

   $$m = \text{———}.$$

   (The letter ____ represents slope.)

   $\dfrac{y_2 - y_1}{x_2 - x_1}$

   m

4. To find the slope of the line through $(-6, 4)$ and $(8, -3)$, let $(x_1, y_1) = (-6, 4)$. Then

   $x_2 =$ ____ and $y_2 =$ ____. The slope is

   $$m = \frac{-3 - \text{\_\_\_}}{8 - \text{\_\_\_}}$$

   $$= \text{\_\_\_}.$$

   8; −3

   4
   (−6)

   $-\dfrac{7}{14}$ or $-\dfrac{1}{2}$

Find the slope of the lines through the following pairs of points.

5. $(-8, 1), (-3, 10)$ $m =$ ____ $\dfrac{9}{5}$

6.3 The Slope of a Line

6. (2, 7), (4, 7) m = _____ | 0

   The line through (2, 7) and (4, 7) is a (*horizontal/vertical*) line. All horizontal lines have a slope of ____. | horizontal

   0

7. (−1, 9), (−1, −2) m = _____ | no slope

   The line through (−1, 9) and (−1, −2) is a (*horizontal/vertical*) line. The slope of all vertical lines is _____. | vertical

   undefined

8. To find the slope of a line from its equation, _____ the equation for ___. | solve; y

9. Let us solve $4x + y = 6$ for y. Subtract ____ from both sides. | 4x

   $4x + y -$ ____ $= 6 - 4x$ | 4x

   $y =$ _____ | 6 − 4x

   The slope of $4x + y = 6$ is given by the number in front of ____ (the _____ of x) after the equation is solved for y. The coefficient of x, and thus the slope, is ____. | x; coefficient

   −4

Find the slope of the lines in Frames 10–12.

10. $7x - 8y = 2$

    Add 8y to both sides, getting

    _____. | 7x = 2 + 8y

    Subtract _____ from both sides. | 2

    _____ $= 8y$ | 7x − 2

    Divide by ____, to get | 8

    _____ $= y$. | $\frac{7}{8}x - \frac{1}{4}$

    The slope is _____. | $\frac{7}{8}$

208   Chapter 6   Graphing Linear Equations

11. $9x + 2y = 5$

            $y =$ _____, slope = _____ $-\frac{9}{2}x + \frac{5}{2}$; $-\frac{9}{2}$

12. $7x + 11y = 4$ slope = _____ $-\frac{7}{11}$

13. Slopes can be used to tell if two lines are _____, perpendicular, or _____. Parallel lines have _____. The slopes of perpendicular lines have a product of ____.

            parallel; neither
            the same slope
            -1

Write *parallel*, *perpendicular*, or *neither* in Frames 14-16.

14. $5x - 2y = 8$
      $2x + 5y = 1$ _____   perpendicular

15. $3x - y = 6$
      $y = 3x - 1$ _____   parallel

16. $6x - y = 4$
      $6x + y = 2$ _____   neither

6.4 Equations of a Line

[1] Write an equation of a line given its slope and y-intercept. (See Frames 1-6 below.)

[2] Graph a line given its slope and a point on the line. (Frames 7-9)

[3] Write an equation of a line given its slope and any point on the line. (Frames 10-14)

[4] Write an equation of a line given two points on the line. (Frames 15-16)

---

1. Recall from the last section that the slope of $y = -5x + 1$ is ____, the coefficient of ____. Note that the equation is solved for ____.

     -5; x
     y

6.4 Equations of a Line

2. In the equation $y = -5x + 1$, we can let $x = 0$ to get

$$y = -5() + 1$$
$$y = \underline{}.$$

	0
	1

Since $y = 1$ when $x = 0$, $(0, 1)$ is the _____ for the graph of $y = -5x + 1$.

y-intercept

3. An equation solved for y, written

$$y = \underline{},$$

is in _____-intercept form, where $(0,)$ is the y-intercept.

	$mx + b$
	slope; b

4. The slope-intercept form can be used to find the equation of a line. For example, the line with slope -3 and y-intercept 7 has equation _____.

$y = -3x + 7$

Find an equation for each line in Frames 5–6.

5. Slope $-\frac{4}{3}$ and y-intercept $\frac{1}{2}$

$$y = \underline{}$$

Eliminate fractions by multiplying both sides by ____ to get

$$\underline{}$$

or $\underline{} = 3$.

	$-\frac{4}{3}x + \frac{1}{2}$
	6
	$6y = -8x + 3$
	$8x + 6y$

6. Slope $\frac{2}{3}$, y-intercept $\frac{5}{12}$

$$\underline{}$$

$12y = 8x + 5$

7. Slope can be used to draw the graph of a line. Suppose a line goes through the point $(3, -2)$ and has slope $-\frac{1}{4}$. First graph the point _____.

$(3, -2)$

210 Chapter 6 Graphing Linear Equations

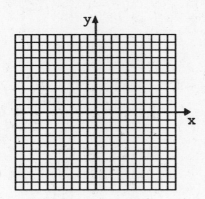

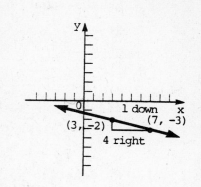

From this point, use the slope, ____, and go ____ unit(s) (*up/down*). Then go ____ unit(s) to the right horizontally. This gives a second point (7, -3). Draw a line through the two points.

$-\dfrac{1}{4}$

1; down; 4

8. To graph the line through (-1, 5) with slope 3/4, graph _____. Then start at this point; go to the right ____ and up ____ to (3, 8). Complete the graph.

(-1, 5)

4; 3

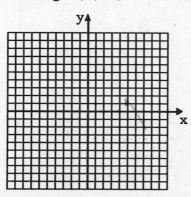

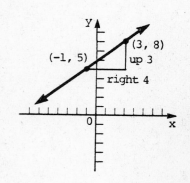

9. Graph the line through (5, 1) with slope -3/2.

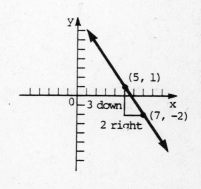

6.4 Equations of a Line

10. Given a point on a line plus the slope of the line, we can find the equation of the line by using the _____ form of the equation of the line. Write the point-slope form.

    $y - y_1 =$ _____

    point-slope

    $m(x - x_1)$

11. To find the equation of the line through (-2, 8), with slope -3, let

    $x_1 =$ ____, $y_1 =$ ____, $m =$ ____.

    -2; 8; -3

    Now use the point-slope form.

    $y - (\quad) = (\quad)(x - \quad)$ 8; -3; -2
    $y - 8 = -3(\quad)$ $x + 2$
    $y - 8 =$ _____ $-3x - 6$
    $y = -3x +$ _____   2

Find an equation of the following lines in the form $Ax + By = C$.

12. Through (1, -3), slope $-\frac{5}{3}$.

    $y - (\quad) = (\quad)(x - \quad)$ $-3; -\frac{5}{3}; 1$
    $y +$ ____ $= -\frac{5}{3}(x - 1)$ 3
    $(\quad)(y + 3) = (\quad)(x - 1)$ 3; -5
    $3y + 9 =$ _____ $-5x + 5$
    $5x + 3y =$ _____ -4

13. Through (6, 5), $m = \frac{3}{4}$ _____ $3x - 4y = -2$

14. Through (2, 7), $m = 0$ _____ $y = 7$

212   Chapter 6 Graphing Linear Equations

15. We can also find an equation of a line if we know two different _____ on the line. For example, to find the equation of the line through (1, -5) and (4, 1), first find the slope of the line, using the definition of slope. The slope is | points

$$m = \frac{-5 - \_\_\_}{1 - \_\_\_}$$

$$= \_\_\_$$

| $\frac{-5 - 1}{1 - 4}$
| 2

Use the slope of _____ and either given point in the _____-slope form of the equation of a line. Let us choose (1, -5). Start with | 2
point

$$y - y_1 = _____.$$

| $m(x - x_1)$

Replace y_1 with ____, m with ____, and x_1 with ____. This gives | -5; 2
1

$$y - (\ \ ) = 2(x - \ \ )$$
$$y + 5 = \_\_\_\_\_,$$
$$y = \_\_\_\_\_$$
$$\_\_\_\_\_ - y = \_\_\_\_\_$$

| -5; 1
| 2x - 2
| 2x - 7
| 2x; 7

16. Find an equation of the line through (2, -1) and (-3, 5).

_____ | $6x + 5y = 7$

6.5 Graphing Linear Inequalities in Two Variables

[1] Graph ≤ or ≥ linear inequalities. (See Frames 1–3 below.)

[2] Graph < or > linear inequalities. (Frames 4–7)

[3] Graph inequalities with a boundary through the origin. (Frames 8–9)

---

1. Earlier in this chapter, you graphed linear equations such as $x + 3y = 8$. In this section, you will graph linear _____, such as $x + y \leq 8$. To graph $x + 3y \leq 8$, first graph the line _____. | inequalities

$x + 3y = 8$

6.5 Graphing Linear Inequalities in Two Variables

Complete the table of values and graph $x + 3y = 8$.

x	y
0	__
__	0
2	__

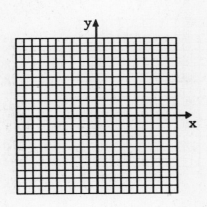

8/3
8
2

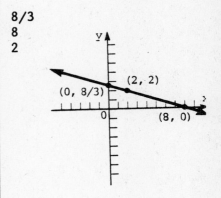

The graph of $x + 3y \leq 8$ will either be the region above the line or the region _____ the line. To decide which, select any point off the line, such as (0, 0). Try (0, 0) in the original equality.

below

$$x + 3y \leq 8$$
$$\underline{} + 3() \leq 8$$
$$\underline{} \leq 8 \quad (true/false)$$

0; 0
0; true

This statement is _____, so you shade the region containing (0, 0). Shade the correct region on the graph above. The final answer is shown at the right.

true

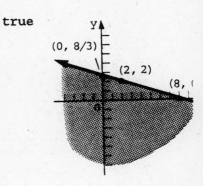

Graph the inequalities in Frames 2 and 3.

2. $x - 2y \leq 6$

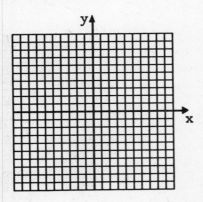

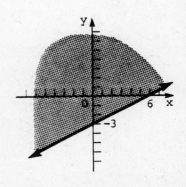

214   Chapter 6 Graphing Linear Equations

3. $x \leq 3$ (Hint: $x = 3$ is a (*vertical/horizontal*) line.)   vertical

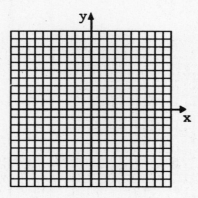

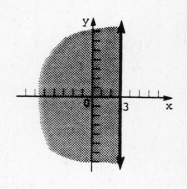

4. To graph $2x - 3y > 6$, first graph _____. Do so on the coordinate axes below. Since the original inequality, $2x - 3y > 6$, does not include the points of the line $2x - 3y = 6$, make the graph of the line _____, and not solid.

$2x - 3y = 6$

dashed

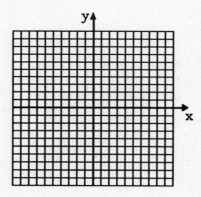

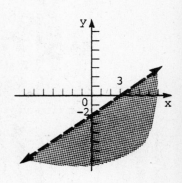

Choose the point $(0, 0)$, a point off the line. Substitute in the original inequality.

$$2x - 3y > 6$$
$$2() - 3() > 6$$ 0; 0
$$\underline{} > 6 \quad (true/false)$$ 0; false

This statement is _____, so shade the region not containing $(0, 0)$. Complete the graph.   false

6.5 Graphing Linear Inequalities in Two Variables

Graph the inequalities in Frames 5–7.

5. $-2x + 3y < 12$

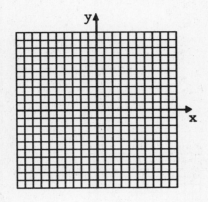

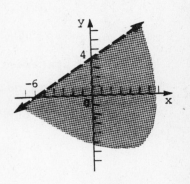

6. $4x - y > 8$

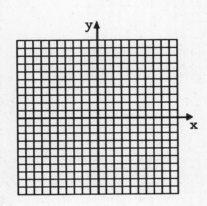

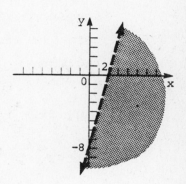

7. $y > -2$

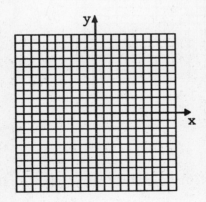

8. To graph $x \leq 3y$, first graph _____. To decide which side of the line to shade, you cannot choose (0, 0), since this point is on the line _____. You must choose a point off the line.

$x = 3y$

$x = 3y$

216   Chapter 6 Graphing Linear Equations

If you choose (4, 0), then

$$x \le 3y$$
$$4 \le 3(\quad)$$
$$4 \le \underline{\qquad}. \quad (true/false)$$

This statement is _____. Thus, you must shade the region not containing (4, 0). Complete the following graph.

0
0; false
false

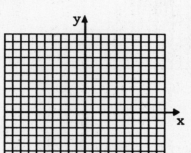
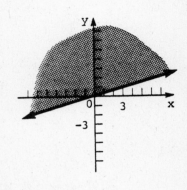

9. Graph $2x + 3y > 0$.

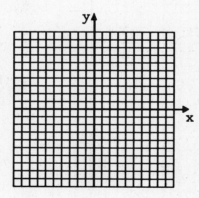
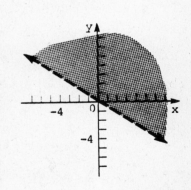

6.6 Functions

[1] Understand the definition of a relation. (See Frames 1–3 below.)

[2] Understand the definition of a function. (Frames 4–6)

[3] Decide whether an equation defines a function. (Frames 7–18)

[4] Find domains and ranges. (Frames 19–22)

[5] Use $f(x)$ notation. (Frames 23–26)

6.6 Functions

1. Any set of ordered pairs is called a _____. | relation
   The set of all first elements in the ordered pairs
   is the _____ of the relation, and the set of | domain
   all second elements in the ordered pairs is the
   _____ of the relation. | range

2. The relation $\{(2, 1), (2, 6), (8, 9)\}$ has domain
   _____ and range _____. | $\{2, 8\}$; $\{1, 6, 9\}$

3. The relation $\{(-1, 4), (1/2, 7), (6, 9), (-2, 9)\}$
   has domain _____ and range _____. | $\{-1, 1/2, 6, -2\}$; $\{4, 7, 9\}$

4. A special type of relation is called a _____. | function
   A function is a relation where each first element
   corresponds to _____ _____ second element. | exactly one

5. The relation in Frame 2 (*is/is not*) a function. | is not
   The number 2 in its domain corresponds to both
   _____ and _____ in its range. | 6; 1

6. The relation in Frame 3 (*is/is not*) a function. | is

7. An equation using x and y can be thought of as a
   _____. For example, y = 2x - 5 is a relation | relation
   between x and y. Elements of this relation in-
   clude (1, -3), (, -5), (-1, -7) and (3,). | 0; 1
   There is an infinite number of elements in this
   relation.

8. Decide whether or not y = 2x - 5 is a function.
   To be a function, there must be exactly one value
   of y for each value of x. If you choose a value
   of x, such as x = 4, then

   $$y = 2x - 5$$
   $$y = 2(\quad) - 5$$ | 4
   $$y = \underline{\quad}.$$ | 3

You found exactly one value of ____ . | y
Choose another value of x, say x = -3.

$$y = 2x - 5$$
$$y = 2() - 5$$ | -3
$$y = $$ | -11

Again, you found exactly ____ value of ____ . | one; y
In fact, for any value of x that you might choose
here, there will be only one value of ____ . | y
Thus, y = 2x - 5 (*is/is not*) a function. | is

9. What about $y = x^2$?

If x = 5, then $y = ()^2 = $ ____ . | $(5)^2$; 25
If x = -5, then $y = ()^2 = $ ____ . | $(-5)^2$; 25
If x = 2, then y = ____ . | 4
If x = -2, then y = ____ . | 4

If you choose any value of x, you will always get
exactly ____ value of y. Therefore, $y = x^2$ | one
(*is/is not*) a _____ . (With this function, | is; function
different x-values may lead to the same y-value,
but there is exactly one value of ____ for each | y
value of ____ .) | x

10. Is $x = y^2$ a function? If x = 16, then y = ____ | 4
or y = ____ . Here, for one value of x we get | -4
____ values of y. Thus, $x = y^2$ (*is/is not*) a | two; is not
function.

11. Is y = 3x - 5 a function? (*yes/no*) | yes

12. Is $y = x^3$ a function? (*yes/not*) | yes

13. Is y = |x| a function? (*yes/not*) | yes
    Recall: |x| is the _____ value of x. | absolute

6.6 Functions 219

14. Is $x = |y|$ a function?
    Notice that if $x = 2$, $y =$ ___ or $y =$ ___.   2; -2
    (yes/no)   no

15. If a _____ line cuts a graph in more than one   vertical
    point, the graph (is/is not) the graph of a   is not
    function. This is called the vertical line_____   test
    for a function.

Write *function* or *not a function* in Frames 16—18.

16.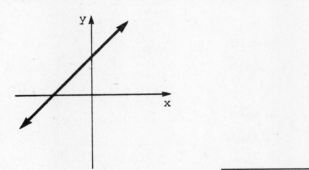

    _____   function

17.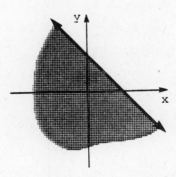

    _____   not a function

18.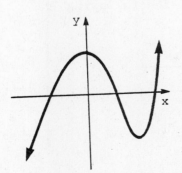

    _____   function

Chapter 6 Graphing Linear Equations

19. For a function, the set of all possible values of x is called the _____ of the function. The set of all possible values of ____ is the _____.

    domain
    y; range

Find the domain and range in Frames 20–22.

20. $x + y = 4$

    Any number can be used for x, so the domain is the set of _____. Any number can be used for y, so the range is also the set of all _____.

    all real numbers
    real numbers

21. $y = x^2 + 6$ Domain: _____   all real numbers
                       Range: _____ $y \geq 6$

22. $y = |x| - 2$ Domain: _____   all real numbers
                       Range: _____ $y \geq -2$

23. Letters like f, g, and h are used to name functions. For example, the function $y = 3x - 5$ is sometimes written as $f(x) = $ _____. If $f(x) = 3x - 5$, then $f(4)$ can be found by substituting ____ for ____.

    $3x - 5$

    4; x

    $f(x) = 3x - 5$
    $f(4) = 3(\quad) - 5$ 4
    $f(4) = $ ____.   7

    $f(4)$ is read "_____."   f of 4

24. Let $f(x) = 3x - 5$, and find each of the following values.

    $f(8) = 3(\quad) - 5 = $ ____   8; 19
    $f(-4) = $ ____ $- 5 = $ ____   −12; −17
    $f(0) = $ ____   −5
    $f(-3) = $ ____   −14

25. Let $g(x) = x^2 + 2$. Then

$g(3) = (\underline{})^2 + 2 = \underline{} + 2 = \underline{}$ 3; 11; 13

$g(-4) = (\underline{})^2 + 2 = \underline{}$ -4; 18

$g(0) = \underline{}$ 2

$g\left(\dfrac{1}{2}\right) = \underline{}$ $\dfrac{9}{4}$

26. Let $P(x) = -x^2 + 6x - 4$. Find the following.

$P(0) = \underline{}$ -4

$P(-3) = -(\underline{})^2 + 6(\underline{}) - 4 = \underline{}$ -3; -3; -31

$P(4) = \underline{}$ 4

Chapter 6 Test

The answers for these questions are at the back of this Study Guide.

Complete the ordered pairs using the given equations.

1. $3x + 5y = 15$ (0,) (, 0) (-5,) (, -3)   1. _____

2. $y = 3x$ (0,) (, 6) (-1,) (, -12)   2. _____

3. $x + 1 = 10$ (, 5) (, 4) (, 0) (, -3)   3. _____

Graph each linear equation. Give the x- and y-intercepts.

4. $x + y = -3$

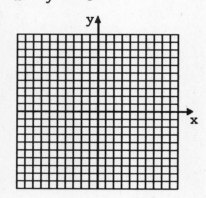

5. $2x - 5y = 10$

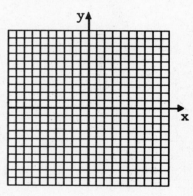

6. $3x = 2y$

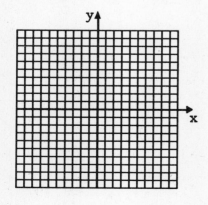

7. $y + 1 = 0$

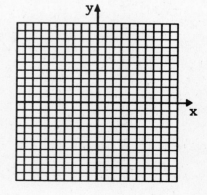

Find the slope of each line.

8. Through (-1, 3) and (2, 7)   8. _____

Chapter 6 Test   223

9. $2x + 11y = 10$ 9. _____

10. $y + 2 = 0$ 10. _____

Write an equation for each line. Write it in the form $Ax + By = C$.

11. Through $(2, -5)$; $m = -2$ 11. _____

12. $m = -1$; y-intercept 4   12. _____

13. Through $(-7, 2)$ and $(1, -4)$ 13. _____

Graph each linear inequality.

14. $3x - y > 6$ 15. $x \geq -4$

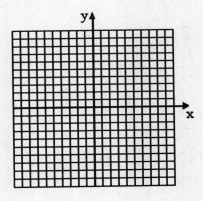   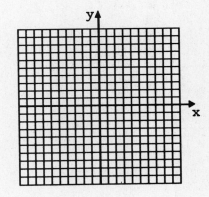

In Problems 16-18, decide which are functions.

16.   16. _____

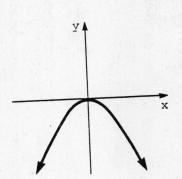

Chapter 6   Graphing Linear Equations

17. $5x + 4 \leq 10$

17. _____

18. $y = 2 + x^2$

18. _____

19. Find the domain and range of $y = -|7 - x|$.

19. _____

20. For $f(x) = -x^2 + 4x - 1$ find $f(-3)$ and $f(5)$.

20. _____

7.1 Solving Systems of Linear Equations by Graphing

CHAPTER 7 LINEAR SYSTEMS

7.1 Solving Systems of Linear Equations by Graphing

[1] Decide whether a given ordered pair is a solution of a system. (See Frames 1-3 below.)

[2] Solve linear systems by graphing. (Frames 4-11)

[3] Identify systems with no solutions or with an infinite number of solutions. (Frames 12-14)

[4] Identify inconsistent systems or systems with dependent equations without graphing. (Frames 15-19)

---

1. A system of equations is two or more _____, each of which contains the same (*variables/terms*). For example, the following is a system of two linear equations:

    $$4x - y = 11$$
    $$2x + 3y = -5.$$

    | equations |
    | variables |

2. To find the solution of a system, you need to find values for each variable that make each _____ true at the same time. For example, is $x = 2$ and $y = -3$ a solution for the system in Frame 1? To find out, substitute ____ for x and ____ for y in each _____, and see if both results are true. First substitute in $4x - y = 11$.

    $$4(\quad) - (\quad) = 11$$
    $$\underline{\quad} = 11 \; (true/false)$$

    Then substitute into $2x + 3y = -5$.

    $$2(\quad) + 3(\quad) = -5$$
    $$\underline{\quad} = -5 \; (true/false)$$

    Therefore, $x = 2$ and $y = -3$ (*is/is not*) the solution of the system

    $$4x - y = 11$$
    $$2x + 3y = -5.$$

    | equation |
    | 2; -3 |
    | equation |
    | 2; -3 |
    | 11; true |
    | 2; -3 |
    | -5; true |
    | is |

226   Chapter 7   Linear Systems

    The solution is often written as the _____   ordered
    pair _____.   (2, -3)

3. Is (-6, 2) a solution for the system

$$x + y = -4$$
$$2x - y = -14? \quad (yes/no)$$
    yes

4. To find the solution of a system of linear equations, draw a _____ of each equation on the   graph
same set of axes. Since the equations of the
system are linear, the graphs are straight
_____. The point of intersection of these   lines
graphs gives the _____ of the system.   solution
Frame 5 gives you an example of the graphing
method of solving a system of linear equations.

5. On the axes at the right, graph $x + y = 5$ and
$x - y = 1$ separately. To graph $x + y = 5$, first
complete the following table of values:

x	y
0	
	0
4	

    5
    5
    1

Then draw a line
through the points
on the graph. To
graph $x - y = 1$,
first complete the
table of values:

x	y
0	
	0
4	

    -1
    1
    3

Now draw the graph.

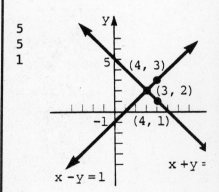

7.1 Solving Systems of Linear Equations by Graphing

The two graphs intersect at the point _____, | (3, 2)
which is the _____ of the system | solution

$$x + y = 5$$
$$x - y = 1.$$

6. Use the graphing method to find the solution of the system

   $$x + 2y = 2$$
   $$x + y = 3.$$

   First complete the table of values for $x + 2y = 2$:

x	y
0	
	0
4	

   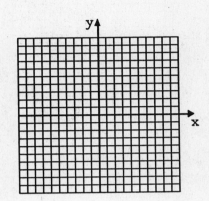

	1
	2
	-1

   Then complete the table of values for $x + y = 3$.

x	y
0	
	0
2	

	3
	3
	1

   The graphs intersect at _____, which is the | (4, -1)
   _____ of the given system. | solution

7. Solve the following system by graphing:

   $$2x + y = -3$$
   $$x - y = -6.$$

228 Chapter 7 Linear Systems

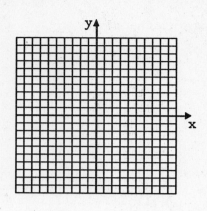

The graphs intersect at the point _____.

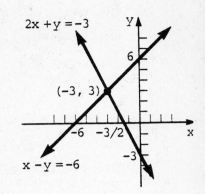

(-3, 3)

8. Solve

   $2x + 3y = -2$
   $3x + 2y = -8$.

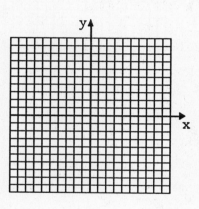

The graphs intersect at _____.

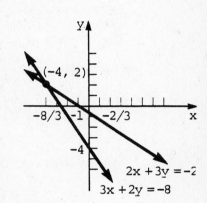

(-4, 2)

9. Solve

   $-3x - 4y = 18$
   $2x + y = -2$

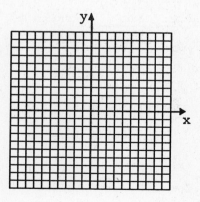

The graphs intersect at _____.

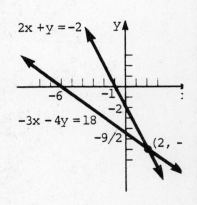

(2, -6)

7.1 Solving Systems of Linear Equations by Graphing   229

10. Solve

$4x - 3y = -12$

$4x + 3y = 24$.

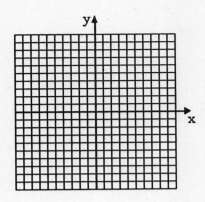
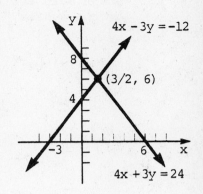

The graphs intersect at _____.   (3/2, 6)

This problem illustrates the difficulty that can
be met to read the _____ of the point   coordinates
that represents the solution. Later in the
chapter algebraic methods are used to find solutions exactly.

11. Thus far, every system we have graphed has a
    solution. A system with a solution is called
    a _____ system.   consistent

12. Graph both lines
    of the system

    $3x + 2y = 6$

    $-6x - 4y = 8$.

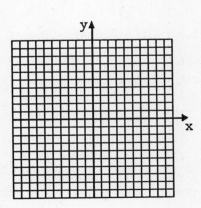
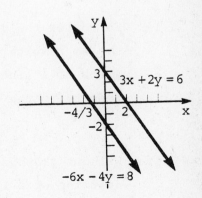

The lines are _____, so they (do/do not)   parallel; do not
intersect. Thus, the system has (one/no) solu-   no
tion. A system of equations with no solutions is
called an _____ system.   inconsistent

230   Chapter 7   Linear Systems

13. Graph both equations
of the system

$x + y = 6$

$-x - y = -6.$

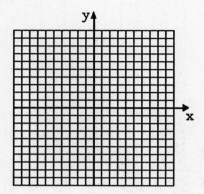

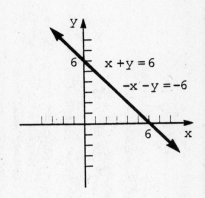

Both equations lead to the _____ line. Thus,   same
there are an infinite number of _____ to   solutions
both equations at the same time. Write this as
"infinite number of solutions" or "_____ line."   same

Because the equations of the system are different
forms of the same equation, these equations are
called _____ equations. On the other hand,   dependent
equations that have different graphs are called
_____ equations.   independent

14. To review the graph of an inconsistent linear
system results in _____ lines. The   parallel
graph of a linear system of dependent equations
results in the _____ _____.   same line

15. Without graphing, it is possible to describe the
solution of a linear system. Consider the system

$$2x + 3y = 6$$
$$4x + 6y = 7.$$

7.1 Solving Systems of Linear Equations by Graphing

Write the slope-intercept form for each equation by solving for _____.	y
$2x + 3y = 6$ $4x + 6y = 7$	
$3y =$ _____ $+ 6$ $6y =$ _____ $+ 7$	$-2x;\ -4x$
$y =$ _____ $y =$ _____	$\frac{-2}{3}x + 2;\ \frac{-2}{3}x + \frac{7}{6}$
Both equations have a slope of _____, but have (*the same/different*) y-intercepts. These equations result in _____ lines if they are graphed. The linear system has ____ solution and is called _____.	$-2/3$ different parallel no inconsistent

16. Consider the system

$$x - 3y = 2$$
$$2x - 6y = 4$$

Write the slope-intercept form for each equation.	
$y =$ _____ $y =$ _____	$\frac{1}{3}x - \frac{2}{3};\ \frac{1}{3}x - \frac{2}{3}$
The equations are (*the same/different*). They are said to be _____ and the system has an _____ number of solutions.	the same dependent infinite

Describe the number of solutions for each system in Frames 17–19.

17. $3x + 2y = 1$
    $x - 2y = 7$ _____   1

18. $2y + 7x = -1$
    $14x - 4y = 1$ _____   none

19. $3y - x = 5$
    $6y = 2x + 10$ _____   infinite number

7.2 Solving Systems of Linear Equations by Addition

[1] Solve linear systems by addition. (See Frames 1-2 below.)

[2] Multiply one or both equations of a system so that the addition method can be used. (Frames 3-10)

[3] Use an alternative method to find the second value in a system. (Frames 11-12)

[4] Use the addition method to solve an inconsistent system. (Frames 13-15)

[5] Use the addition method to solve a system of dependent equations. (Frames 16-22)

---

1. To solve the system

    $$x + y = 12$$
    $$x - y = 2$$

    by the addition method, draw a line under the two equations, and _____ them. | add

    $$x + y = 12$$
    $$\underline{x - y = 2}$$
    Sum: _____ | $2x + 0 = 14$ or $2x = 14$

    If $2x = 14$, then $x =$ _____. To find y, substitute _____ for x in either _____ of the system. Do this in the first equation. | 7

    7; equation

    $$x + y = 12$$
    $$\underline{\quad} + y = 12$$
    $$y = \underline{\quad}$$ | 7

    5

    The solution of the system is _____. Check by substituting _____ for x and _____ for y in the second equation. | (7, 5)

    7; 5

    $$x - y = 2$$
    $$(\ ) - (\ ) = 2 \quad (true/false)$$ | 7; 5; true

7.2 Solving Systems of Linear Equations by Addition

2. Solve by the addition method.

$$x + y = 5$$
$$3x - y = 7$$

Draw a line under the two equations and _____ them. Which variable will be eliminated? _____

| add |
| y |

$$x + y = 5$$
$$3x - y = 7$$

Sum: _____

$4x = 12$

Then $x =$ _____. Substitute _____ for x in the first equation, and find y.

3; 3

$$x + y = 5$$
$$\underline{} + y = 5$$
$$y = \underline{}$$

3

2

The solution of the system is _____.

(3, 2)

3. Solve $x + 2y = 9$
$3x - y = -8$.

If you add these equations together, you (*would/ would not*) get an equation with only one variable. But you want to eliminate one variable when the equations are added. To do that, multiply the top equation by _____. This gives

would not

−3

( )x + ( )$2y$ = ( )9

or _____.

−3; −3; −3

$-3x - 6y = -27$

Now add this equation and the second equation of the system.

$$-3x - 6y = -27$$
$$\underline{3x - y = -8}$$

Sum: _____

$-7y = -35$

234 Chapter 7 Linear Systems

Then y = _____. Substitute _____ for y in either | 5; 5
of the original equations of the system. If you
choose the first equation, then

$$x + 2() = 9$$ | 5
$$x = \underline{}$$ | −1

The solution of the system is _____. | (−1, 5)

4. Solve by multiplying the top equation by 3:

$$3x - y = 17$$
$$2x + 3y = -7.$$

Then x = _____. Substitute _____ for x in either | 4; 4
of the two original equations, and find y:
y = _____, so that the solution of the system is | −5
_____. | (4, −5)

5. Solve $3x + y = 3$
$4x + 3y = -1.$

Multiply the top equation by _____, and add. | −3
New equation: | −9x − 3y = −9

$$\underline{4x + 3y = -1}$$
Sum: _____ | −5x = −10

x = _____. Use this value of x to find y. | 2
The solution of the system is _____. | (2, −3)

6. $2x + y = -1$
$5x + 4y = 8$ Solution: _____ | (−4, 7)

7. $2x + 3y = 2$
$3x - 2y = 16$

Here it is necessary to multiply _____ equa- | both
tions. Multiply the top equation by _____, and | 3

7.2 Solving Systems of Linear Equations by Addition

multiply the bottom equation by ____. Do this, and continue the solution.

-2

$$6x + 9y = 6$$
$$\underline{-6x + 4y = -32}$$
$$13y = -26$$

y = ____. Now find x.

-2

The solution of the system is _____.

$(4, -2)$

8. $3x - 4y = 6$
    $4x - 5y = 9$

    Multiply the top equation by 4 and the bottom equation by ____. Then add.

-3

$$12x - 16y = 24$$
$$\underline{-12x + 15y = -27}$$
$$-y = -3$$

From your last equation, y = ____. Find x.

3

     Solution: _____

$(6, 3)$

9. $4x + 3y = 8$
    $6x + 5y = 10$ Solution: _____

$(5, -4)$

10. $2x + 4y = 7$
    $4x - 6y = -7$

$-4x - 8y = -14$
$\underline{4x - 6y = -7}$
$-14y = -21$
$y = 3/2$

$4x - 6(3/2) = -7$
$4x = -7 + 9$
$x = 2/4$
$= 1/2$

     Solution: _____

$(1/2, 3/2)$

11. To solve the system

    $2x = -1 - 4y$
    $8y = 8x - 11$,

236 Chapter 7 Linear Systems

first rewrite the system above, getting

        _____   2x + 4y = −1
        _____   8x − 8y = 11

Solve this system.

$$\begin{aligned} 4x + 8y &= -2 \\ 8x - 8y &= 11 \\ \hline 12x &= 9 \end{aligned}$$

                                          x = _____   3/4

In this case, it might be easier to find the value of the second variable by using the addition method _____.   twice

Multiply the first equation by _____ and solve for y.   −4

$$\begin{aligned} -8x - 16y &= 4 \\ 8x - 8y &= 11 \\ \hline -24y &= 15 \end{aligned}$$

                                          y = _____   −5/8

The solution is _____.   (3/4, −5/8)

12.   Use the alternative method to solve the system

$$x = 2y + 2$$
$$12y = 13 - 9x.$$

First rewrite the system.

$$\begin{aligned} x - 2y &= 2 \\ 9x + 12y &= 13 \end{aligned}$$

$$\begin{aligned} 6x - 12y &= 12 \\ 9x + 12y &= 13 \\ \hline 15x &= 25 \end{aligned}$$

                                          x = _____   5/3

Multiply the first equation by _____.   −9

$$\begin{aligned} -9x + 18y &= -18 \\ 9x + 12y &= 13 \\ \hline 30y &= -5 \end{aligned}$$

                                          y = _____   −1/6

The solution is _____.   (5/3, −1/6)

7.2 Solving Systems of Linear Equations by Addition

13. The graph of one linear equation is a _____ line. If there are two linear equations in a system, then the graphs of these two equations can intersect in one of three possible ways. | straight

(a) The two lines may intersect in exactly _____ point. In this case, the solution of the system is given by the _____ of the _____ where the lines interect. This is the _____ of the system. Examples are in Section 7.1. | one

coordinates

point

solution

(b) The two lines may be parallel. In this case, the lines (do/do not) intersect, so that there is _____ solution for the system. See Frames 14 and 15 below for an example. These systems are called _____. | do not

no

inconsistent

(c) The two lines may be the _____ line, that is, the two lines coincide. In this case, there are an _____ number of solutions. You can describe this situation by writing "_____ _____ of _____." See Frames 16 and 17 below for an example. These systems have equations that are _____. | same

infinite

infinite
number; solutions

dependent

14. Graph the lines of the system

$x + y = 2$
$x + y = 6$.

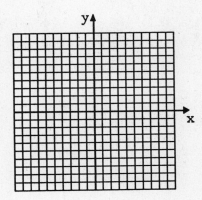
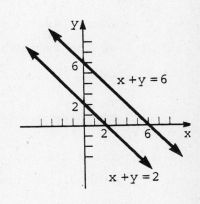

238   Chapter 7 Linear Systems

These lines are _____, so that there is _____ solution for the system.	parallel no
15. Solve the system in Frame 14 by the addition method. Multiply the top equation by _____, and add. New equation: $$x + y = 6$$ Sum: _____ (*true/false*)	-1 -x - y = -2 0 = 4; false
The statement 0 = 4 is _____. This means that the given system contradicts _____. It has _____ solution, the graphs of the equations are _____ lines. The system is _____.	false itself no parallel; inconsistent
16. Graph the lines of the system $$2x + 2y = 8$$ $$x + y = 4.$$ 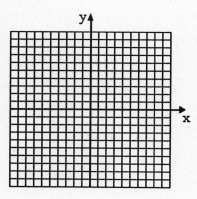	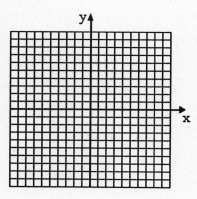
These lines are the _____ line, so the solution of the system can be described by saying _____ _____ The equations are _____.	same "infinite number of solutions." dependent
17. Solve the system in Frame 16 by the addition method. Multiply the second equation by ____, and add.	-2

7.2 Solving Systems of Linear Equations by Addition

$$2x + 2y = 8$$

New equation: _____ | $-2x - 2y = -8$

Sum: _____ (*true/false*) | $0 = 0$; true

The statement $0 = 0$ is ____. This means that every solution of one equation (*is/is not*) a solution of the other. | true
 | is

The system has (*no/an infinite number of*) solution(s). The equations of this system are _____. | an infinite number of
 | dependent

In Frames 18–22, use the addition method on each system of equations, and decide which kind of solution applies.

18. $4x - 2y = 6$
    $2x - y = 4$

    Multiply the second equation by ____, then add. | -2

    $$4x - 2y = 6$$
    _____ | $-4x + 2y = -8$

    Sum: _____ (*true/false*) | $0 = -2$; false

    The lines are _____, and there is ____ solution for the system. | parallel; no

19. $3x - 5y = 9$
    $6x - 10y = 18$

    | $-6x + 10y = -18$
    | $\underline{6x - 10y = 18}$
    | $0 = 0$;
    | true

    The lines are the _____ line. There are (*an infinite number of/no*) solutions. | same
     | an infinite number of

240 Chapter 7 Linear Systems

20. $\frac{2}{3}x - \frac{4}{3}y = \frac{8}{3}$
    $2x - 4y = 8$

	$-2x + 4y = -8$
	$\underline{2x - 4y = 8}$
	$0 = 0;$ true
Solution: _____	same line, or an infinite number of solutions

21. $\frac{5}{9}x - \frac{4}{3}y = 1$
    $10x - 24y = 24$

	$-10x + 24y = -18$
	$\underline{10x - 24y = 24}$
	$0 = 6;$ false
Solution: _____	no solution

22. $5x = 3y - 8$
    $10x = 9 + 6y$

Solution: _____   no solution

7.3 Solving Systems of Linear Equations by Substitution

[1] Solve linear systems by substitution. (See Frames 1–8 below.)

[2] Solve linear systems with fractions. (Frames 9–10)

---

1. Sometimes when you are given a system of equations to solve, one of the equations is already solved for one of the variables. An example of such a system is

   $2x + y = 5$
   $x = y + 4.$

   To solve this system, we can substitute $y + 4$ for _____ in the first equation.   x

   $2(\quad) + y = 5$ $y + 4$

7.3 Solving Systems of Linear Equations by Substitution

Simplify this equation.

$2y + \underline{} + y = 5$	8
$\underline{} + 8 = 5$	$3y$
$3y = \underline{}$	-3
$y = \underline{}$	-1

Substitute _____ for y in the (*first/second*) equation. | -1; second

$x = \underline{} + 4$	-1
$x = \underline{}$	3

The solution of the system is _____. | $(3, -1)$

Solve the systems in Frames 2–6.

2. $3x - 2y = 8$
   $x = y + 2$

   Substitute _____ for _____ in the _____ equation. | $y + 2$; x; first

$3() - 2y = 8$	$y + 2$
$\underline{} - 2y = 8$	$3y + 6$
$y = \underline{}$	2

   Now find _____. Since $x = y + 2$, | x

$x = () + 2$	2
$x = \underline{}$.	4

   The solution of the system is _____. | $(4, 2)$

3. $4x + 3y = 5$
   $x = 2y - 7$

$4() + 3y = 5$	$2y - 7$
$\underline{} + 3y = 5$	$8y - 28$
$11y = \underline{}$	33
$y = \underline{}$	3

Chapter 7 Linear Systems

Now find x.

$$x = 2() - 7 \qquad 3$$
$$x = \underline{} \qquad -1$$

The solution of the system is _____. (-1, 3)

4. $3x + 5y = 19$
   $x + 2y = 8$

Before you can use the substitution method here, first rewrite the second equation so that it is solved for x. To solve $x + 2y = 8$ for x, subtract ____ from both sides. Doing this, you get   2y

$$x = \underline{}. \qquad 8 - 2y$$

Now substitute in the first equation.

$$3x + 5y = 19$$
$$3() + 5y = 19 \qquad 8 - 2y$$
$$\underline{} + 5y = 19 \qquad 24 - 6y$$
$$-y = \underline{} \qquad -5$$
$$y = \underline{} \qquad 5$$

Use this value of y to find x.
The solution of the system is _____. (-2, 5)

5. $5x + 2y = 4$
   $3x + y = 4$

Solve the second equation for ____. (Solve for   y
y since solving for x would result in fractions.)

$$y = \underline{} \qquad 4 - 3x$$

Now substitute in the first equation, and complete the solution.

The solution of the system is _____. (4, -8)

6. $3x + 4y + 11 = 0$
   $2x + y = 1$

Solution: _____ (3, -5)

7.3 Solving Systems of Linear Equations by Substitution

7. Solve the system

$$2y = 25 - 4x$$
$$2x + y = 8$$

by substitution.

Solve $2x + y = 8$ for y.

$$y = \underline{\hspace{1cm}} \qquad\qquad 8 - 2x$$

Substitute $8 - 2x$ for y in the equation $2y = 25 - 4x$.

$$2(\quad) = 25 - 4x \qquad\qquad 8 - 2x$$
$$\underline{\hspace{1cm}} = 25 - 4x \qquad\qquad 16 - 4x$$

Add 4x on both sides.

$$\underline{\hspace{1cm}} = 25 \qquad\qquad 16$$

This result is (*true/false*). Because of this,   false
the system has ____ solution.   no

8. Solve the system

$$x + y = 6$$
$$3x = 18 - 3y$$

by substitution.

Solve $x + y = 6$ for x. (We also could have solved for y.)

$$x = \underline{\hspace{1cm}} \qquad\qquad 6 - y$$

Substitute _____ for x in the equation $3x = 18 - 3y$. $6 - y$

$$3(\quad) = 18 - 3y \qquad\qquad 6 - y$$
$$\underline{\hspace{1cm}} = 18 - 3y \qquad\qquad 18 - 3y$$

Add 3y on both sides.

$$\underline{\hspace{1cm}} = 18 \qquad\qquad 18$$

This result is (*true/false*) so that the system   true
has equations that represent the ____ ____.   same line

244 Chapter 7 Linear Systems

9. To solve the system

$$\frac{x}{2} + \frac{y}{4} = 1$$
$$\frac{x}{4} + y = -3,$$

clear of fractions by multiplying the top equation on both sides by _____. | 4

$$4\left(\frac{x}{2} + \frac{y}{4}\right) = 4()$$ | 1
$$\underline{} = 4$$ | $2x + y$

Multiply both sides of the second equation by _____. | 4

This gives

$$\underline{} = -12.$$ | $x + 4y$

We now have the system

$$\underline{}$$ | $2x + y = 4$
$$\underline{}.$$ | $x + 4y = -12$

Solve the first of these equations for y.

$$y = \underline{}$$ | $4 - 2x$

Substitute _____ for y in the second equation. | $4 - 2x$

$$x + 4() = -12$$ | $4 - 2x$
$$x + \underline{} = -12$$ | $16 - 8x$
$$\underline{} = -12$$ | $16 - 7x$
$$-7x = \underline{}$$ | -28
$$x = \underline{}$$ | 4

Since $y = 4 - 2x$, we have

$$y = 4 - 2()$$ | 4
$$y = \underline{}.$$ | -4

The solution of the system is _____. | $(4, -4)$

10. Solve the system

$$\frac{2x}{3} + \frac{y}{6} = 9$$
$$\frac{x}{4} - \frac{y}{3} = 1.$$

Solution: _____ | (12, 6)

7.4 Applications of Linear Systems

[1] Use linear systems to solve problems about numbers. (See Frames 1-2 below.)

[2] Use linear systems to solve problems about quantities and their costs. (Frames 3-4)

[3] Use linear systems to solve problems about mixtures. (Frames 5-6)

[4] Use linear systems to solve problems about rate or speed using the distance formula. (Frames 7-8)

---

1. **The sum of two numbers is 15. Twice the first is three more than the second. Find the two numbers.**

   Here we let x = the first number and _____ = the second number. | y

   Write "the sum of two numbers is 15" as an equation. _____ | x + y = 15

   Write "twice the first is three more than the second" as an equation. That is, "twice _____ is three more than _____." The equation is _____. | x
   y
   2x = y + 3

   Using these equations, write the system. | x + y = 15
   2x = y + 3

   Solve the first equation for y: y = _____. Substitute this into the second equation, and solve the system. | 15 - x

   x = _____ | 6
   y = _____ | 9

Chapter 7 Linear Systems

The two numbers are ____. | 6 and 9

Check: ____ + ____ = 15 (*true/false*) | 6; 9; true

2( ) = ____ + 3 (*true/false*) | 6; 9; true

2. The difference between two numbers is 8. Twice the smaller is 2 less than the larger. Find the two numbers.

   Let x = the larger number and y = the smaller number.

   "The difference between two number is 8":

   _____ | $x - y = 8$

   "Twice the smaller is 2 less than the larger":

   _____ | $2y = x - 2$

   Write the system of equations, and solve it.

   | $x - y = 8$
   | $2y = x - 2$
   | $x = 14; y = 6$

   The two numbers are _____. | 14 and 6

3. In total, 200 adult and student tickets were sold for the school play. Adult tickets cost $5 and student tickets cost $3. A total of $840 was received from ticket sales. Find the number sold of each kind of ticket.

   Let x = the number of adult tickets and y = the number of student tickets. Complete the following chart.

Kind of tickets	Number sold	Cost of each (in dollars)	Receipts (in dollars)
Adult	x	5		5x
Child	y	3		3y

7.4 Applications of Linear Systems 247

We must use the information in the chart to find two equations. Since 200 tickets were sold,

$$\underline{} = 200.$$ $x + y$

How much money in total was received for the tickets? _____ This money comes from two parts: adult tickets and child tickets. This produces the equation $840

$$\underline{} = 840.$$ $5x + 3y$

Solve the system consisting of this equation and the equation _____. $x + y = 200$

     Number of adult tickets = ____   120
     Number of child tickets = ____   80

4. Harriet bought a total of 170 $.19 stamps and $.29 stamps, paying $43.30 for them. Find the number of each kind of stamp she bought.

Let x = the number of $.19 stamps and y = the number of $.29 stamps.

     The equation for the total number is _____. $x + y = 170$
     The equation for the total value is _____. $.19x + .29y = 43.30$

To clear the second equation of decimals, multiply both sides by ____.   100

$$\underline{} = 4330$$ $19x + 29y$

Solve the system.

                       $x =$ ____   60
                       $y =$ ____   110

Harriet bought ____ $.19 stamps and ____ $.29 stamps.   60; 110

Chapter 7 Linear Systems

5. A candy merchant plans to mix some candy worth $2.50 a pound with some $4.00 candy to get 30 pounds of mixture worth $3.50 a pound. Find the number of pounds of each type of candy she should use.

   Let x = the number of pounds of $2.50 candy and y = the number of pounds of $4.00 candy. Complete this chart.

Price per pound	Number of pounds	Value
2.50	x	
4.00	y	
3.50	30	

   2.50x
   4.00y
   3.50(3)

   From the middle column, we get the equation

   _____ .

   x + y = 30

   From the right-hand column, we get

   _____ .

   2.50x + 4.00y
   = 3.50(30)

   Since 3.50(30) = ____, this second equation is

   _____ = 105.

   105

   2.50x + 4.00y

   Solve the system.

   x = _____
   y = _____

   10
   20

   She should use _____ of $2.50 candy and _____ of $4.00 candy.

   10 pounds
   20 pounds

6. Some 80% antifreeze and some 40% antifreeze are mixed to form 100 quarts of a 56% mixture. Find the number of quarts of each type that are used.

   Let x = the number of quarts of 80% antifreeze,
   y = the number of quarts of 40 antifreeze.

7.4 Applications of Linear Systems 249

The equation for total quantity is _____.	x + y = 100
The equation for total antifreeze is _____.	.80x + .40y = 56

Solve the system.

Number of quarts of 80% = _____	40
Number of quarts of 40% = _____	60

7. Two cars start at the same point and travel in opposite directions. One car travels 40 kilometers per hour faster than the other. After 6 hours, they are 960 kilometers apart. Find the speed of each car.

   Let x = the speed of the slower car and y = the speed of the faster car. Complete this chart. (Use the formula d = _____.)

   rt

	r	t	d
Slower car	x	6	
Faster car	y	6	

   6x

   6y

   Write an equation for the speeds of the cars.

   _____

   x + 40 = y or
   y − 40 = x

   Write an equation for the distance.

   6x + 6y = _____

   960

   Solve the system.

speed of slower car = _____	60 kilometers per hour
speed of faster car = _____	100 kilometers per hour

8. A boat travels 15 miles per hour with the current and 9 miles per hour against the current. Find the speed of the boat in still water and the speed of the current.

250   Chapter 7   Linear Systems

speed of boat = _____	12 miles per hour
speed of current = _____	3 miles per hour

7.5 Solving Systems of Linear Inequalities

[1] Solve systems of linear inequalities by graphing.
(See Frames 1-4 below.)

Before starting Frame 1, review the ways of graphing an inequality in two variables, such as $3x + 4y \geq 12$. See Section 6.5 in the previous chapter.

---

1. To graph a system of inequalities such as

   $$x + y \leq 6$$
   $$2x - 3y \leq 6,$$

   graph each inequality separately on the same axes. The solution of the system is given by the overlap or _____ of the two graphs. Graph each inequality of the system. Then shade the area of intersection heavily enough so that it will stand out.

   intersection

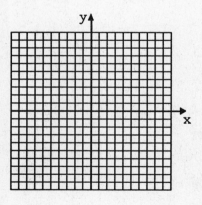
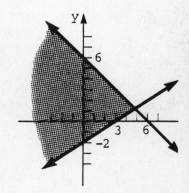

7.5 Solving Systems of Linear Inequalities   251

2. Graph the system
   $x + 2y \leq 6$
   $y \geq 2x - 8$.

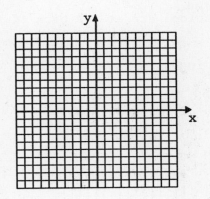
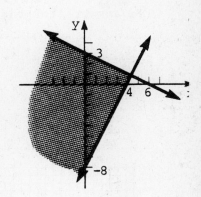

3. Graph the system
   $3x - 2y > 12$
   $4x + 3y < 12$.

   (Don't forget to make the boundary lines _____.)

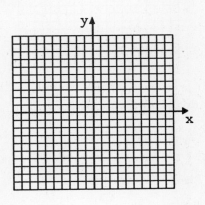
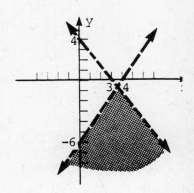

4. $x > y + 4$
   $y \geq -2$

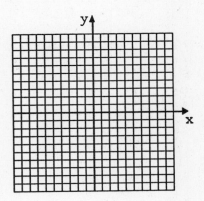
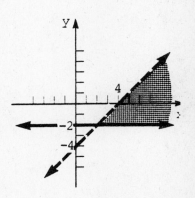

Chapter 7 Test

The answers for these questions are at the back of this Study Guide.

Solve each system by graphing.

1. $4x + y = 8$
   $2x - y = -2$

2. $x + 3y = 6$
   $x - y = 2$

3. $4x + 7y = 3$
   $8x + 14y = 5$

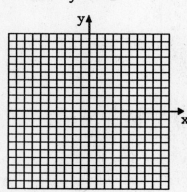
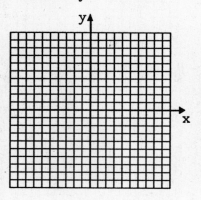
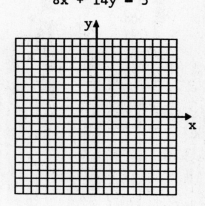

Solve each system by the addition method.

4. $2x - 3y = 7$
   $5x + 3y = 7$

   4. _____

5. $4x + 5y = 2$
   $3x + 3y = 5$

   5. _____

6. $2x + 7y = -13$
   $\frac{3x}{2} + y = 3$

   6. _____

7. $5x - 3y = -26$
   $6x + 5y = 29$

   7. _____

8. $7x - 2y = 1$
   $14x - 4y = -2$

   8. _____

9. $\frac{2x}{3} - \frac{5y}{8} = -22$
   $\frac{5x}{9} + \frac{3y}{4} = 2$

   9. _____

Solve each system by substitution.

10. $4x - 3y = 44$
     $x = -2y$

11. $2x + 5y = 36$
     $5x - y = 9$

Solve each system by any method.

12. $7 + 9x - 3y = 8y - 39$
     $2x + 7y + 4 = 9x + 3y + 17$

13. $\dfrac{5x}{2} + \dfrac{y}{3} = -12$
     $\dfrac{3x}{2} - \dfrac{2y}{9} = -11$

Solve each problem by a system of equations.

14. The sum of two numbers is 25. If one number is doubled, it equals 1 less than the other. Find the numbers.

15. The local shoe store is having a sale. Some shoes cost $23 a pair, and some cost $28 a pair. Mike has exactly $125 to spend and wants to buy 5 pairs of shoes. How many can he buy at each price?

16. A 70% solution of acid is to be mixed with a 40% solution to get 120 liters of a 50% solution. How many liters of each solution should be used?

10. _____

11. _____

12. _____

13. _____

14. _____

15. _____

16. _____

254 Chapter 7 Linear Systems

17. A boat can go 45 miles upstream in 3 hours, while it takes 4 hours to go 100 miles downstream. Find the speed of the current and the speed of the boat in still water.

17. _____

Graph the solution of each system of inequalities.

18. $3x + 5y \leq 15$
    $x + y \geq 4$

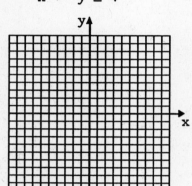

19. $3x - 2y < 6$
    $4x + y < 4$

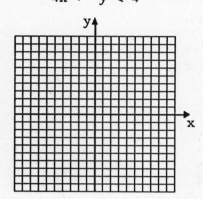

20. $5x + 2y \geq 10$
    $x \leq 4$

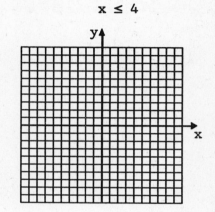

CHAPTER 8 ROOTS AND RADICALS

8.1 Finding Roots

$\boxed{1}$ Find square roots. (See Frames 1–5 below.)

$\boxed{2}$ Decide whether a given root is rational, irrational, or not a real number. (Frames 6–9)

$\boxed{3}$ Find decimal approximations for irrational square roots. (Frame 10)

$\boxed{4}$ Use the Pythagorean theorem. (Frames 11–13)

$\boxed{5}$ Find higher roots. (Frames 14–17)

---

1. To square 4, multiply 4 by itself:

    $4 \cdot 4 = 4^{\underline{}} = \underline{}$. | 2; 16

   The reverse process is to take the square _____ | root
   of 16. The number 16 has two square roots, 4 and
   -4, since $4^2 = 16$ and also $()^2 = 16$. | -4

2. Find all the square roots of the following numbers.

100 ___ and ___	$\frac{1}{4}$ ___ and ___	10; -10; $\frac{1}{2}$; $-\frac{1}{2}$
81 ___ and ___	$\frac{9}{16}$ ___ and ___	9; -9; $\frac{3}{4}$; $-\frac{3}{4}$
25 ___ and ___	$\frac{144}{25}$ ___ and ___	5; -5; $\frac{12}{5}$; $-\frac{12}{5}$
144 ___ and ___	1 ___ and ___	12; -12; 1; -1
225 ___ and ___	0 ___	15; -15; 0

3. In symbols, the positive square root of 16 is

   $\sqrt{16} = 4$. The expression $\sqrt{16}$ is called a _____. | radical

   The $\sqrt{}$ is a _____ sign, and the number under | radical
   the radical sign is called the _____. When a | radicand
   radical sign is used for square root, you always
   take the _____ square root, for example | positive
   $\sqrt{64} = \underline{}$, $\sqrt{9} = \underline{}$, and so no. Note that | 8; 3

256   Chapter 8   Roots and Radicals

$-\sqrt{9} = -(\ \ ) =$ _____. If the words "square roots" are used, as in Frame 2, then both the positive and _____ roots are meant.

3; −3

negative

4. For the following radicals, fill in the blank.

$\sqrt{49} =$ _____ $\sqrt{169} =$ _____

$\sqrt{\dfrac{100}{169}} =$ _____ $\sqrt{\dfrac{9}{25}} =$ _____

$\sqrt{121} =$ _____ $\sqrt{1} =$ _____

$\sqrt{0} =$ _____

7; 13

$\dfrac{10}{13}$; $\dfrac{3}{5}$

11; 1

0

5. Since 4 is a rational number, and $4^2 = 16$, the number 16 is a perfect square. Then 100, 81, 9/4, 25, 144, and 36/49 are all _____ squares. A perfect square is the square of a _____ number. Thus

$\sqrt{100} =$ _____, $\sqrt{81} =$ _____, $\sqrt{9/4} =$ _____,

$\sqrt{25} =$ _____, $\sqrt{144} =$ _____, and $\sqrt{36/49} =$ _____.

perfect

rational

10; 9; 3/2

5; 12; 6/7

6. Find the square roots of the number 7. The number 7 has _____ square roots, one positive and one negative. However, there is no _____ number which is a square root of 7. Thus the square roots of 7 must be shown by a _____ sign as

_____ and _____.

Since $\sqrt{7}$ is not a rational number, it is called an _____ number.

two

rational

radical

$\sqrt{7}$; $-\sqrt{7}$

irrational

7. Write the two square roots of 40:

_____ and _____.

Is $\sqrt{40}$ irrational? (yes/no)

$\sqrt{40}$; $-\sqrt{40}$

yes

8.1 Finding Roots   257

8. Write the square roots of 85: _____ $\sqrt{85}$ and $-\sqrt{85}$

   Is $-\sqrt{85}$ irrational? (yes/no)   yes

9. If a number is _____, then no real number   negative

   is its square root. For example, $\sqrt{-49}$ (is/is not)   is not

   a real number.

10. You can find decimal approximations for irrational
    numbers by using the square root key of a calcu-
    lator. Use a calculator to find decimal approxi-
    mations for the following. The symbol $\approx$ means
    "is _____ equal to." Round your   approximately
    answers to the nearest thousandths.

    $\sqrt{11} \approx$ _____ $\sqrt{20} \approx$ _____   3.317; 4.472
    $\sqrt{39} \approx$ _____ $\sqrt{80} \approx$ _____   6.245; 8.944

11. Square roots can be used with the Pythagorean
    formula. In the right triangle shown,

    $a^2 +$ ____ = ____ . b^2; c^2

    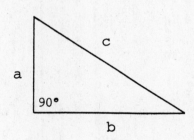

    The _____ of the triangle are represented by a   legs
    and b. The _____ is represented by c.   hypotenuse

12. Given two sides of a right triangle, we can use
    the Pythagorean formula to find the length of the
    third side. For example, if a = 5 and b = 6, then

    $a^2 + b^2 =$ _____ c^2

becomes ( )² + ( )² = c², | 5; 6

or ____ + ____ = c² | 25; 36

____ = c². | 61

From this result,

$$c = \underline{}.$$

$\sqrt{61}$

(Use only the _____ square root.) | positive

Using a calculator gives

$$\sqrt{61} \approx \underline{}.$$

7.810

13. In a right triangle, a = 9 and c = 13. Find b.

$$b = \sqrt{\underline{}} \approx \underline{}$$

88; 9.381

14. Since $2^3 = 2 \cdot 2 \cdot 2 = 8$, then 2 is the _____ | cube

root of 8. In symbols, $\sqrt[3]{8} = \underline{}$. Cube roots | 2

of negative numbers are real numbers, for example,

$\sqrt[3]{-8} = \underline{}$. | -2

Find each of the following roots.

15. $\sqrt[3]{64} = \underline{}$ | 4

16. $\sqrt[4]{16} = \underline{}$ | 2

17. $\sqrt[5]{-32} = \underline{}$ | -2

8.2 Multiplication and Division of Radicals

[1] Multiply radicals. (See Frames 1-2 below.)

[2] Simplify radicals using the product rule. (Frames 3-12)

[3] Simplify radical quotients using the quotient rule. (Frames 13-22)

[4] Use the product and quotient rules to simplify higher roots. (Frames 23-26)

---

1. If a and b are positive numbers, then

   $$\sqrt{a} \cdot \sqrt{b} = \sqrt{}.$$ ab

   This result is called the _____ rule for rad-   product
   icals. For example,

   $$\sqrt{8} \cdot \sqrt{32} = \sqrt{8()} = \sqrt{} = \underline{}.$$ 32; 256; 16

2. Also, $\sqrt{5} \cdot \sqrt{8} = \sqrt{}$ and $\sqrt{15} \cdot \sqrt{3} = \sqrt{}$.   40; 45

3. You can use the product rule to simplify radicals.
   For example, you can simplify $\sqrt{28}$ by finding the
   largest perfect _____ factor of 28. (A perfect   square
   square is the square of a _____ number.)   rational
   The largest perfect square factor of 28 is ___.   4
   Therefore,

   $$\sqrt{28} = \sqrt{4()}$$ 7

   $$= \sqrt{} \cdot \sqrt{}$$ 4; 7

   $$= \underline{\phantom{2\sqrt{7}}}.$$ $2\sqrt{7}$

Simplify each radical in Frames 4-12.

4. $\sqrt{8} = \sqrt{()()} = \sqrt{} \cdot \sqrt{}$ 4; 2; 4; 2

   $= \underline{\phantom{2\sqrt{2}}}$ $2\sqrt{2}$

5. $\sqrt{27} = \sqrt{()()} = \sqrt{} \cdot \sqrt{}$ 9; 3; 9; 3

   $= \underline{\phantom{3\sqrt{3}}}$ $3\sqrt{3}$

6. $\sqrt{45} = \sqrt{()()} = \sqrt{} \cdot \sqrt{}$ | 9; 5; 9; 5
   $= \underline{\phantom{3\sqrt{5}}}$ | $3\sqrt{5}$

7. $\sqrt{54} = \sqrt{()()} = \sqrt{} \cdot \sqrt{}$ | 9; 6; 9; 6
   $= \underline{\phantom{3\sqrt{6}}}$ | $3\sqrt{6}$

8. $\sqrt{200} = \sqrt{()()} = \underline{\phantom{10\sqrt{2}}}$ | 100; 2; $10\sqrt{2}$

9. $\sqrt{72} = \sqrt{()()} = \underline{\phantom{6\sqrt{2}}}$ | 36; 2; $6\sqrt{2}$

10. $\sqrt{128} = \sqrt{()()} = \underline{\phantom{8\sqrt{2}}}$ | 64; 2; $8\sqrt{2}$

11. $\sqrt{108} = \sqrt{()()} = \underline{\phantom{6\sqrt{3}}}$ | 36; 3; $6\sqrt{3}$

12. $\sqrt{98} = \sqrt{()()} = \underline{\phantom{7\sqrt{2}}}$ | 49; 2; $7\sqrt{2}$

13. If a and b are positive numbers, then

$$\sqrt{\frac{a}{b}} = \underline{}.$$

| $\dfrac{\sqrt{a}}{\sqrt{b}}$

This result is called the quotient rule for radicals. For example,

$$\sqrt{\frac{25}{36}} = \underline{} = \underline{}.$$

| $\dfrac{\sqrt{25}}{\sqrt{36}}$; $\dfrac{5}{6}$

Simplify each radical in Frames 14–26. Assume all variables are positiave.

14. $\sqrt{\dfrac{8}{9}} = \underline{}$ Quotient rule | $\dfrac{\sqrt{8}}{\sqrt{9}}$

   $= \underline{}$ 9 is a perfect square | $\dfrac{\sqrt{8}}{3}$

   $= \underline{}$ | $\dfrac{2\sqrt{2}}{3}$

15. $\sqrt{\dfrac{40}{9}} =$ _____ | $\dfrac{2\sqrt{10}}{3}$

16. $\dfrac{\sqrt{40}}{\sqrt{10}} = \sqrt{} = \sqrt{} =$ ____ | $\dfrac{40}{10}$; 4; 2

17. $\dfrac{\sqrt{80}}{\sqrt{5}} =$ ____ | 4

18. $\dfrac{9\sqrt{14}}{3\sqrt{2}} =$ ____ | $3\sqrt{7}$

19. $\sqrt{\dfrac{7}{8}} \cdot \sqrt{24} =$ ____ | $\sqrt{21}$

20. $\sqrt{49y^8} =$ ____ | $7y^4$

21. $\sqrt{z^{15}} = \sqrt{z^{14} \cdot } =$ ____ $\sqrt{z}$ | z; z^7

22. $\sqrt{\dfrac{50}{p^2}} =$ ____ | $\dfrac{5\sqrt{2}}{p}$

23. $\sqrt[3]{24} = \sqrt[3]{ \cdot 3} =$ ____ $\sqrt[3]{3}$ | 8; 2

24. $\sqrt[3]{\dfrac{16}{27}} =$ ____ | $\dfrac{2\sqrt[3]{2}}{3}$

25. $\sqrt[3]{432} = \sqrt[3]{ \cdot 2} =$ ____ $\sqrt[3]{2}$ | 216; 6

26. $\sqrt[4]{48} = \sqrt[4]{ \cdot 3} = \sqrt[4]{} \cdot \sqrt[4]{3} =$ ____ | 16; 16; $2\sqrt[4]{3}$

262   Chapter 8   Roots and Radicals

8.3   Addition and Subtraction of Radicals

[1]   Add and subtract radicals.   (See Frames 1-4 below.)

[2]   Simplify radical sums and differences.   (Frames 5-10)

[3]   Simplify radical expressions involving multiplication.   (Frames 11-12)

---

1. Earlier, we defined like terms as terms with exactly the same _____ raised to the same _____. We saw then that only _____ terms can be added or subtracted. In the same way, like radicals have exactly the same number under the radical sign, or like radicals have the same _____. For example, $2\sqrt{7}$ and $-5\sqrt{7}$ are _____ radicals, as are $\sqrt{5}$, $-3\sqrt{\ \ }$, and $8\sqrt{\ \ }$. Are $-5\sqrt{7}$ and $\sqrt{5}$ like radicals? (yes/no) Are $3\sqrt{3}$ and $2\sqrt{3}$ like radicals? (yes/no)

   variables

   powers; like

   radicands; like

   5; 5

   no

   yes

2. Only _____ terms can be added and subtracted. In a similar way, only _____ radicals can be _____ and subtracted. For example, by the _____ property,

   $$3\sqrt{5} + 7\sqrt{5} = (\ \ )\sqrt{5} = \_\_\_\_\_.$$
   Also, $8\sqrt{2} - 12\sqrt{2} = (\ \ )\sqrt{2} = \_\_\_\_\_.$

   like

   like; added

   distributive

   3 + 7; $10\sqrt{5}$

   8 - 12; $-4\sqrt{2}$

3. The sum $\sqrt{5} + 5\sqrt{2}$ (can/cannot) be further simplified.

   cannot

4. $4\sqrt{2} - 6\sqrt{2} + 11\sqrt{2} = (\ \ \ \ )\sqrt{2} = \_\_\_\_\_$

   4 - 6 + 11; $9\sqrt{2}$

5. Some problems require that we first simplify each radical in a sum or difference. For example, to simplify $3\sqrt{50} - 4\sqrt{2}$, first simplify _____.

   $\sqrt{50} = \sqrt{(\ \ )(\ \ )} = \_\_\_\_\_$

   $\sqrt{50}$

   25; 2; $5\sqrt{2}$

8.3 Addition and Subtraction of Radicals

Thus, the original problem becomes

$3\sqrt{50} - 4\sqrt{2} = 3(\quad) - 4\sqrt{2}$ $5\sqrt{2}$

$= \underline{\quad\quad} - 4\sqrt{2}$ $15\sqrt{2}$

$= \underline{\quad\quad}.$ $11\sqrt{2}$

Simplify in Frames 6–10.

6. $6\sqrt{8} + 2\sqrt{18} - 5\sqrt{72}$

   Simplify $\sqrt{8}$, $\sqrt{18}$, and $\sqrt{72}$ in the first step.

   $6\sqrt{8} + 2\sqrt{18} - 5\sqrt{72} = 6(\quad) + 2(\quad) - 5(\quad)$ $2\sqrt{2};\ 3\sqrt{2};\ 6\sqrt{2}$

   $= \underline{\quad} + \underline{\quad} - \underline{\quad}$ $12\sqrt{2};\ 6\sqrt{2};\ 30\sqrt{2}$

   $= \underline{\quad\quad}$ $-12\sqrt{2}$

7. $-3\sqrt{20} + 2\sqrt{45} - 3\sqrt{125}$

   $= -3(\quad) + 2(\quad) - 3(\quad)$ $2\sqrt{5};\ 3\sqrt{5};\ 5\sqrt{5}$

   $= \underline{\quad\quad}$ $-15\sqrt{5}$

8. $-6\sqrt{24} + 2\sqrt{150} = -6(\quad) + 2(\quad) = \underline{\quad\quad}$ $2\sqrt{6};\ 5\sqrt{6};\ -2\sqrt{6}$

9. $\dfrac{2\sqrt{3}}{3} + \dfrac{2\sqrt{12}}{3} - \dfrac{5\sqrt{27}}{3}$

   $= \dfrac{2\sqrt{3}}{3} + \dfrac{2(\quad)}{3} - \dfrac{5(\quad)}{3}$ $2\sqrt{3};\ 3\sqrt{3}$

   $= \underline{\quad} + \underline{\quad} - \underline{\quad}$ $\dfrac{2\sqrt{3}}{3};\ \dfrac{4\sqrt{3}}{3};\ \dfrac{15\sqrt{3}}{3}$

   $= \dfrac{\underline{\quad\quad}}{3}$ $2\sqrt{3} + 4\sqrt{3} - 15\sqrt{3}$

   $= \underline{\quad\quad}$ $-3\sqrt{3}$

264   Chapter 8 Roots and Radicals

10. $5\sqrt[3]{24} - 7\sqrt[3]{3} = 5\sqrt[3]{\cdot 3} - 7\sqrt[3]{3}$	8
$= 5\sqrt[3]{} \cdot \sqrt[3]{3} - 7\sqrt[3]{3}$	8
$= 5()\sqrt[3]{3} - 7\sqrt[3]{3}$	2
$= \underline{\phantom{10\sqrt[3]{3}}} - 7\sqrt[3]{3}$	$10\sqrt[3]{3}$
$= \underline{\phantom{3\sqrt[3]{3}}}$	$3\sqrt[3]{3}$
11. To simplify $\sqrt{8} \cdot \sqrt{16} - 4\sqrt{2}$, we must first _____.	multiply
$\sqrt{8} \cdot \sqrt{16} - 4\sqrt{2} = \sqrt{} - 4\sqrt{2}$	128
$= \underline{\phantom{8\sqrt{2}}} - 4\sqrt{2} = \underline{\phantom{4\sqrt{2}}}$	$8\sqrt{2}$; $4\sqrt{2}$
12. Simplify $\sqrt{5} \cdot \sqrt{10p} + \sqrt{8p}$. _____	$7\sqrt{2p}$

8.4 Rationalizing the Denominator

1️⃣ Rationalize denominators with square roots. (See Frames 1–10 below.)

2️⃣ Write radicals in simplified form. (Frames 11–13)

3️⃣ Rationalize denominators with cube roots. (Frames 14)

1. The expression $2/\sqrt{5}$ contains a radical in the denominator. Expressions of this type are difficult to work with. To eliminate the radical in the denominator, multiply numerator and _____ by	denominator
_____. This gives	$\sqrt{5}$
$\dfrac{2 \cdot (\phantom{\sqrt{5}})}{\sqrt{5}(\phantom{\sqrt{5}})} = \dfrac{2\sqrt{5}}{}.$	$\dfrac{\sqrt{5}}{\sqrt{5}}$; 5
(By the product rule, $\sqrt{5} \cdot \sqrt{5} = \sqrt{} = \underline{}$.)	25; 5
This process is called _____ the denominator.	rationalizing

8.4 Rationalizing the Denominator

Rationalize the denominator in Frames 2–10.

2. $\dfrac{5}{\sqrt{6}} = \dfrac{5(\quad)}{\sqrt{6}(\quad)} = \underline{\qquad}$ $\dfrac{\sqrt{6}}{\sqrt{6}};\ \dfrac{5\sqrt{6}}{6}$

3. $\dfrac{8}{\sqrt{7}} = \dfrac{8(\quad)}{\sqrt{7}(\quad)} = \underline{\qquad}$ $\dfrac{\sqrt{7}}{\sqrt{7}};\ \dfrac{8\sqrt{7}}{7}$

4. $\dfrac{16}{\sqrt{2}} = \dfrac{16(\quad)}{\sqrt{2}(\quad)}$ $\dfrac{\sqrt{2}}{\sqrt{2}}$

   $= \dfrac{16\sqrt{2}}{\underline{\quad}} = \underline{\qquad}$, after simplifying. $2;\ 8\sqrt{2}$

5. $\dfrac{27}{\sqrt{3}} = \underline{\qquad} = \underline{\qquad}$ $\dfrac{27(\sqrt{3})}{\sqrt{3}(\sqrt{3})};\ 9\sqrt{3}$

6. $\dfrac{48}{\sqrt{8}}$

   $\sqrt{8}$ can be simplified $\sqrt{(\quad)(\quad)} = \underline{\qquad}$. $4;\ 2;\ 2\sqrt{2}$

   $\dfrac{48}{\sqrt{8}} = \dfrac{48}{2\sqrt{2}} = \dfrac{\underline{\quad}}{\sqrt{2}}$ *Lowest terms*   24

   $= \underline{\qquad}$ *Rationalize the denominator* $\dfrac{24(\sqrt{2})}{\sqrt{2}(\sqrt{2})}$

   $= \underline{\qquad}$ *Product rule* $\dfrac{24\sqrt{2}}{2}$

   $= \underline{\qquad}$ *Lowest terms* $12\sqrt{2}$

7. $\dfrac{9}{\sqrt{12}} = \underline{\qquad}$ *Simplify $\sqrt{12}$* $\dfrac{9}{2\sqrt{3}}$

   $= \underline{\qquad}$ *Rationalize the denominator* $\dfrac{9(\sqrt{3})}{2\sqrt{3}(\sqrt{3})}$

   $= \underline{\qquad}$ *Product rule* $\dfrac{9\sqrt{3}}{2 \cdot 3}$

   $= \underline{\qquad}$ *Lowest terms* $\dfrac{3\sqrt{3}}{2}$

Chapter 8 Roots and Radicals

8. $\dfrac{9}{\sqrt{18}}$ = _____ Simplify $\sqrt{18}$ $\dfrac{9}{3\sqrt{2}}$

    = _____ Lowest terms $\dfrac{3}{\sqrt{2}}$

    = _____ Rationalize the denominator $\dfrac{3\sqrt{2}}{2}$

9. $\dfrac{16}{\sqrt{32}}$ = _____ Simplify $\sqrt{32}$ $\dfrac{16}{4\sqrt{2}}$

    = _____ Lowest terms $\dfrac{4}{\sqrt{2}}$

    = _____ Rationalize the denominator $2\sqrt{2}$

10. $\dfrac{15}{\sqrt{50}}$ = _____ $\dfrac{3\sqrt{2}}{2}$

Write the following radicals in simplified form.

11. $\sqrt{\dfrac{5}{8}}$ = _____ Quotient rule $\dfrac{\sqrt{5}}{\sqrt{8}}$

    = _____ Simplify $\sqrt{8}$ $\dfrac{\sqrt{5}}{2\sqrt{2}}$

    = _____ Rationalize the denominator $\dfrac{\sqrt{5}(\sqrt{2})}{2\sqrt{2}(\sqrt{2})}$

    = _____ Product rule $\dfrac{\sqrt{10}}{2 \cdot 2}$

    = _____ Multiply $\dfrac{\sqrt{10}}{4}$

12. $\sqrt{\dfrac{7}{27}}$ = _____ $\dfrac{\sqrt{21}}{9}$

13. $\sqrt{\dfrac{16p}{q}}$ = _____ $\dfrac{4\sqrt{pq}}{q}$

14. Simplify $\sqrt[3]{\frac{1}{4}}$.

$$\sqrt[3]{\frac{1}{4}} = \frac{1}{\underline{}}$$

$\sqrt[3]{4}$

Multiply numerator and denominator by $\sqrt[3]{2}$, since

$\sqrt[3]{4} \cdot \sqrt[3]{2} = \sqrt[3]{\underline{}} = \underline{}$.

8; 2

$\sqrt[3]{\frac{1}{4}} = \underline{}$

$\frac{\sqrt[3]{2}}{2}$

8.5 Simplifying Radical Expressions

[1] Simplify radical expressions with sums. (See Frames 1-2 below.)

[2] Simplify radical expressions with products. (Frames 3-12)

[3] Simplify radical expressions with quotients. (Frames 13-16)

[4] Write radical expressions with quotients in lowest terms. (Frames 17-20)

1. To simplify a radical, use the rules found in Section 8.5 of the textbook. For example, sums or differences with _____ radicals must be combined. Thus,

   $\sqrt{75} + \sqrt{48} = \underline{} + \underline{} = \underline{}$.

   like

   $5\sqrt{3}$; $4\sqrt{3}$; $9\sqrt{3}$

2. Simplify $8\sqrt{3} + 2\sqrt{27} - 5\sqrt{12}$. _____

   $4\sqrt{3}$

3. Products of radicals are found in much the same way that we multiplied expressions containing variables. For example, the product $\sqrt{2}(8 - 3\sqrt{2})$ is found as follows.

   $\sqrt{2}(8 - 3\sqrt{2}) = ()() - ()()$

   $= 8\sqrt{2} - 3()()$

   $= 8\sqrt{2} - 3()$

   $= 8\sqrt{2} - \underline{}$

   $\sqrt{2}$; 8; $\sqrt{2}$; $3\sqrt{2}$

   $\sqrt{2}$; $\sqrt{2}$

   2

   6

Chapter 8 Roots and Radicals

(Remember: $\sqrt{2} \cdot \sqrt{2} = \sqrt{} = \underline{}$. In general, | 4; 2
$\sqrt{a} \cdot \sqrt{a} = \underline{}$, for any positive a.) | a

4. To multiply radicals such as $(\sqrt{3} - 4)(2\sqrt{3} + 5)$, follow the patterns for multiplying polynomials.

   $(\sqrt{3} - 4)(2\sqrt{3} + 5)$

   $= ()() + (\sqrt{3})(5) - (4)(2\sqrt{3}) - 4()$ | $\sqrt{3}$; $2\sqrt{3}$; 5
                                                  FOIL

   $= 2() + 5\sqrt{3} - 8\sqrt{3} - \underline{}$ *Multiply* | 3; 20

   $= \underline{}$ | $-14 - 3\sqrt{3}$

5. The product $(3 + \sqrt{2})(5 - \sqrt{2})$ is found as follows.

   $(3 + \sqrt{2})(5 - \sqrt{2})$

   $= ()() + 3() + \sqrt{2}() + (\sqrt{2})()$ | 3; 5; $-\sqrt{2}$; 5; $-\sqrt{2}$

   $= 15 - 3\sqrt{2} + \underline{} - \underline{}$ | $5\sqrt{2}$; 2

   $= \underline{}$ | $13 + 2\sqrt{2}$

Find each of the products of Frames 6–12.

6. $\sqrt{8}(3\sqrt{2} - 2) = 3\sqrt{2}() - 2()$ | $\sqrt{8}$; $\sqrt{8}$

                  $= 3() - 2()$ | $\sqrt{16}$ or 4; $2\sqrt{2}$

                  $= \underline{}$ | $12 - 4\sqrt{2}$

7. $(3\sqrt{2} + \sqrt{5})(4\sqrt{2} - 3\sqrt{5})$

   $= (3\sqrt{2})() + (3\sqrt{2})(-3\sqrt{5}) + (\sqrt{5})(4\sqrt{2})$ | $4\sqrt{2}$

       $+ ()()$ | $\sqrt{5}$; $-3\sqrt{5}$

   $= \underline{} - 9\sqrt{10} + 4\sqrt{10} - \underline{}$ | 24; 15

   $= \underline{}$ | $9 - 5\sqrt{10}$

8.5 Simplifying Radical Expressions

8. $(2\sqrt{5} - 3)(3\sqrt{5} + 4)$ _____	$18 - \sqrt{5}$
9. $(2\sqrt{3} - 5\sqrt{2})(\sqrt{3} + 2\sqrt{2})$ _____	$-14 - \sqrt{6}$
10. $(2 + 3\sqrt{2})^2 = ($ ____ $)($ ____ $)$ _____	$2 + 3\sqrt{2};\ 2 + 3\sqrt{2}$ $22 + 12\sqrt{2}$
11. $(3 + \sqrt{5})(3 - \sqrt{5}) = ($ ____ $)^2 - ($ ____ $)^2$ $= $ ____ $-$ ____ $=$ ____	$3;\ \sqrt{5}$ $9;\ 5;\ 4$
12. $(2\sqrt{3} - 3\sqrt{5})(2\sqrt{3} + 3\sqrt{5})$ _____	-33
13. To simplify the expression $$\frac{2}{3 + \sqrt{2}},$$ multiply numerator and denominator by _____, which is called the _____ of $3 + \sqrt{2}$. $$\frac{2}{3 + \sqrt{2}} = \frac{2(\quad)}{(3 + \sqrt{2})(3 - \sqrt{2})}$$ $$= \frac{}{(\quad)^2 - (\quad)^2}$$ $$= \frac{6 - 2\sqrt{2}}{}$$ This process is called _____ the denominator.	$3 - \sqrt{2}$ conjugate $3 - \sqrt{2}$ $6 - 2\sqrt{2}$ $3;\ \sqrt{2}$ 7 rationalizing

Chapter 8 Roots and Radicals

Use conjugates to rationalize the denominators in Frames 14–16.

14. $\dfrac{-3}{2 - \sqrt{5}} = \dfrac{-3()}{(2 - \sqrt{5})()}$ | $2 + \sqrt{5}$
$2 + \sqrt{5}$

$= \dfrac{}{()^2 - ()^2}$ | $-6 - 3\sqrt{5}$
$2; \sqrt{5}$

$= \dfrac{-6 - 3\sqrt{5}}{}$ | -1

$= \dfrac{-1()}{-1}$ | $6 + 3\sqrt{5}$

$= \underline{}$ | $6 + 3\sqrt{5}$

15. $\dfrac{5}{4 - \sqrt{12}} = \dfrac{5()}{(4 - \sqrt{12})()}$ | $4 + \sqrt{12}$
$4 + \sqrt{12}$

$= \dfrac{}{()^2 - ()^2}$ | $20 + 5\sqrt{12}$
$4; \sqrt{12}$

$= \dfrac{20 + 5\sqrt{12}}{}$ | 4

$= \dfrac{20 + 5()}{4}$ | $2\sqrt{3}$

$= \dfrac{20 + }{4}$ | $10\sqrt{3}$

$= \dfrac{2()}{4}$ | $10 + 5\sqrt{3}$

$= \underline{}$ | $\dfrac{10 + 5\sqrt{3}}{2}$

16. $\dfrac{2}{3 - \sqrt{2}}$

$\underline{}$ | $\dfrac{6 + 2\sqrt{2}}{7}$

17. To write $(6 + 4\sqrt{3})/2$ in lowest terms, _____ the numerator. The _____ factor of the numerator and denominator is ___. | factor
common
2

$\dfrac{6 + 4\sqrt{3}}{2} = \dfrac{2()}{2}$ | $3 + 2\sqrt{3}$

$= \underline{}$ | $3 + 2\sqrt{3}$

18. $\dfrac{5\sqrt{7} + 10}{10} =$ _____ | $\dfrac{\sqrt{7} + 2}{2}$

19. $\dfrac{4\sqrt{3} - 2}{6} =$ _____ | $\dfrac{2\sqrt{3} - 1}{3}$

20. $\dfrac{-4 + 16\sqrt{2}}{4} =$ _____ | $-1 + 4\sqrt{2}$

8.6 Equations with Radicals

[1] Solve equations with radicals. (See Frames 1–6 below.)

[2] Identify equations with no solutions. (Frames 7)

[3] Solve equations that require squaring a binomial. (Frames 8–10)

---

1. To solve an equation involving a radical, such as $\sqrt{2x - 1} = 5$, we can use the _____ property of equality. According to the squaring property of equality, if $a = b$, then _____. In words: we can _____ both sides of an _____. All potential solutions from the squared equation must be checked in the _____ equation.

   squaring

   $a^2 = b^2$

   square; equation

   original

2. To solve $\sqrt{2x - 1} = 5$, square both sides.

   $(\quad\quad)^2 = (\quad\quad)^2$

   _____ = _____

   $x =$ _____

   Check this answer:

   $\sqrt{2x - 1} = 5$

   $\sqrt{2(\quad) - 1} = 5$

   $\sqrt{} = 5$

   _____ = 5. (true/false)

   $\sqrt{2x - 1}$; 5

   $2x - 1$; 25

   13

   13

   25

   5; true

Chapter 8 Roots and Radicals

3. To solve the equation $\sqrt{5x - 6} = 8$, square both sides.

$$(\underline{})^2 = (\underline{})^2$$

$$\underline{} = \underline{}$$

$$x = \underline{}$$

Check this answer.

| $\sqrt{5x - 6}$; 8 |
| $5x - 6$; 64 |
| 14 |

Solve the equations in Frames 4–7. Check each answer.

4. $\sqrt{x - 9} = 3\sqrt{2}$

$$(\underline{})^2 = (\underline{})^2$$

$$\underline{} = \underline{}$$

$$x = \underline{}$$

| $\sqrt{x - 9}$; $3\sqrt{2}$ |
| $x - 9$; 18 |
| 27 |

5. $2\sqrt{r} = \sqrt{3r + 9}$

$$(\underline{})^2 = (\underline{})^2$$

$$\underline{} = \underline{}$$

$$r = \underline{}$$

| $2\sqrt{r}$; $\sqrt{3r + 9}$ |
| $4r$; $3r + 9$ |
| 9 |

6. $2\sqrt{m + 3} = \sqrt{5m + 6}$

$$m = \underline{}$$

| 6 |

7. $\sqrt{x + 11} = -4$

$$(\underline{})^2 = (\underline{})^2$$

$$\underline{} = \underline{}$$

$$x = \underline{}$$

Check this proposed answer.

$$\sqrt{x + 11} = -4$$

$$\sqrt{\underline{} + 11} = -4$$

$$\sqrt{\underline{}} = -4$$

$$\underline{} = -4 \quad (true/false)$$

| $\sqrt{x + 11}$; -4 |
| $x + 11$; 16 |
| 5 |
| |
| |
| 5 |
| 16 |
| 4; false |

8.6 Equations with Radicals

This last result is a _____ statement. Therefore, 5 (*is/is not*) a solution of the given equation. There (*is a/is no*) solution.	false is not is no

8. Some equations require that we square a binomial. Remember:

$$(a + b)^2 = \underline{\qquad}.$$

$a^2 + 2ab + b^2$

Use this to solve $\sqrt{x + 1} + 2 = x - 3$.

First, subtract _____ from both sides.   2

$$\underline{\qquad} = \underline{\qquad}$$

$\sqrt{x + 1}$; $x - 5$

Square both sides.

$$x + 1 = \underline{\qquad}$$

$x^2 - 10x + 25$

Make the right side 0.

$$\underline{\qquad} = 0$$

$x^2 - 11x + 24$

Factor.

$$(\quad)(\quad) = 0$$

$x - 8$; $x - 3$

Set each factor equal to 0.

$x - 8 = 0$ or $\underline{\qquad} = 0$ $x - 3$

Solve each equation.

$x = \underline{\qquad}$ or $x = \underline{\qquad}$ 8; 3

Let $x = 8$.   Let $x = 3$.

$\sqrt{x + 1} + 2 = x - 3$ $\sqrt{x + 1} + 2 = x - 3$

$\sqrt{\underline{\ } + 1} + 2 = \underline{\ } - 3$ $\sqrt{\underline{\ } + 1} + 2 = \underline{\ } - 3$ 8; 8; 3; 3

$\underline{\ } = \underline{\ }$ $\underline{\ } = \underline{\ }$ 5; 5; 4; 0

(*true/false*)   (*true/false*)   true; false

The only solution is _____.   8

Chapter 8 Roots and Radicals

9. $\sqrt{x+2} = x - 4$

   $()^2 = ()^2$ $\sqrt{x+2}$; $x-4$

   $\underline{} = \underline{}$ $x+2$; $x^2-8x+16$

   $\underline{} = 0$ $x^2 - 9x + 14$

   $()() = 0$ $x-7$; $x-2$

   $x = \underline{}$ or $x = \underline{}$ 7; 2

   Check both answers.

   Let $x = 7$.   Let $x = 2$.

   $\sqrt{x+2} = x - 4$ $\sqrt{x+2} = x - 4$

   $\sqrt{\underline{}+2} = \underline{} - 4$ $\sqrt{\underline{}+2} = \underline{} - 4$ 7; 7; 2; 2

   $\underline{} = \underline{}$ $\underline{} = \underline{}$ 3; 3; 2; −2

   (*true/false*)   (*true/false*)   true; false

   The only solution is $\underline{}$.   7

10. Solve $\sqrt{m} + 3 = 4m$.

    Subtract $\underline{}$ from both sides.   3

    $\sqrt{m} = \underline{}$ $4m - 3$

    Square both sides.

    $m = \underline{}$ $16m^2 - 24m + 9$

    Subtract m from both sides.

    $\underline{} = 16m^2 - \underline{} + 9$ 0; 25m

    Factor.

    $0 = ()()$ $m-1$; $16m-9$

    Place each factor equal to 0, and solve.

    $m - 1 = 0$ or $\underline{} = 0$ $16m - 9$

    $m = \underline{}$ or $m = \underline{}$ 1; 9/16

    Is $m = 1$ a solution? (*yes/no*)   yes

    Is $m = 9/16$ a solution? (*yes/no*)   no

    The only solution is $\underline{}$.   1

8.7 Fractional Exponents

$\boxed{1}$ Define and use $a^{1/n}$. (See Frames 1-6 below.)

$\boxed{2}$ Define and use $a^{m/n}$. (Frames 7-16)

$\boxed{3}$ Use rules for exponents with fractional exponents. (Frames 17-20)

$\boxed{4}$ Use fractional exponents to simplify radicals. (Frames 21-24)

---

1. Now we study fractional _____. By definition,

    $a^{1/n} = $ _____,

    for a nonnegative number a and a positive _____ n.

    | exponents
    | $\sqrt[n]{a}$
    | integer

2. For example, $49^{1/2} = $ ____, or just ____. | $\sqrt{49}$; 7

Simplify each expression in Frames 3-6.

3. $81^{1/2} = $ ____ | 9

4. $8^{1/3} = $ ____ | 2

5. $16^{1/4} = $ ____ | 2

6. $32^{1/5} = $ ____ | 2

7. By definition,

    $a^{m/n} = $ ____ or ____. | $\sqrt[n]{a^m}$; $(\sqrt[n]{a})^m$

8. For example,

    $8^{2/3} = (\sqrt[3]{})^2$ | 8
    $= ()^2$ | 2
    $= $ ____. | 4

Chapter 8 Roots and Radicals

Simplify each expression in Frames 9–12.

9. $100^{3/2} = (\sqrt{})^3 = ()^3 = \underline{}$ 100; 10; 1000

10. $32^{4/5} = (\sqrt[5]{})^4 = \underline{}$ 32; 16

11. $64^{4/3} = \underline{}$ 256

12. $(-8)^{4/3} = (\sqrt[3]{})^4 = \underline{}$ -8; 16

13. Evaluate negative fractional exponents using

$$a^{-m/n} = \underline{}. \quad (n \underline{} 0)$$

     $1/a^{m/n}$; >

14. For example,

$$8^{-4/3} = \frac{1}{\underline{}}$$

$$= \underline{}.$$

     $8^{4/3}$

     $\dfrac{1}{16}$

Evaluate each expression in Frames 15–16.

15. $125^{-2/3} = \underline{}$ $\dfrac{1}{25}$

16. $1000^{-4/3} = \underline{}$ $\dfrac{1}{10,000}$

Use the properties of exponents in Frames 17–20.

17. $9^{3/2} \cdot 9^{-5/2} = 9^{\underline{}+\underline{}}$

            $= \underline{} = \underline{}$

     3/2; (-5/2)

     9^{-1}; $\dfrac{1}{9}$

18. $(25^{1/4})^2 = 25^{\underline{}}$

            $= \underline{}$

     1/2

     5

8.7 Fractional Exponents

19. $\dfrac{100^{-1/4}}{100^{1/4}} = 100^{-1/4-\underline{}}$ 1/4

     $= 100^{\underline{}}$ $-1/2$

     $= \underline{}$ $\dfrac{1}{100^{1/2}}$ or $\dfrac{1}{10}$

20. $\dfrac{5^{-1/2} \cdot 5^{-1}}{5^{-5/2}} = \underline{}$ 5

21. Some radical expressions can be simplified by writing the expression with _____.   exponents

Simplify each radical expression by writing in expponential form.

22. $\sqrt[4]{16^2} = (16^2)^{\underline{}} = 16^{\underline{}}$ 1/4; 2/4 or 1/2

     $= \underline{}$ $\sqrt{16}$ or 4

23. $\sqrt[3]{5^6} = 5^{\underline{}} = 5^{\underline{}}$ 6/3; 2

     $= \underline{}$ 25

24. $\sqrt[6]{a^3} = a^{\underline{}} = a^{\underline{}}$ 3/6; 1/2

     $= \underline{}$ (if $a \geq 0$) $\sqrt{a}$

Chapter 8 Roots and Radicals

Chapter 8 Test

The answers for these questions are at the back of this Study Guide. In this test, assume that all variables represent positive real numbers.

Find the indicated root.

1. $\sqrt{121}$

2. $-\sqrt{3600}$

3. $\sqrt[3]{-125}$

Simplify where possible.

4. $\sqrt{48}$

5. $\sqrt{\dfrac{108}{49}}$

6. $\dfrac{56\sqrt{48}}{7\sqrt{3}}$

7. $\sqrt[3]{-16}$

8. $\sqrt{20} - \sqrt{45}$

9. $5\sqrt{20} + 7\sqrt{75}$

10. $a\sqrt{32} - \sqrt{8a^2}$

11. $4\sqrt{45p} - 2\sqrt{20p} + 6\sqrt{125p}$

12. $\sqrt{625a^5 b^{10}}$

1. _____

2. _____

3. _____

4. _____

5. _____

6. _____

7. _____

8. _____

9. _____

10. _____

11. _____

12. _____

13. $(9 - \sqrt{2})(9 + \sqrt{2})$

14. $(5 + 3\sqrt{5})(2 - \sqrt{5})$

15. $(\sqrt{7} + \sqrt{3})^2$

Rationalize each denominator.

16. $\dfrac{5\sqrt{8}}{\sqrt{10}}$

17. $\dfrac{4z}{\sqrt{q}}$

18. $\sqrt{\dfrac{7}{8}}$

19. $\dfrac{-7}{1 + \sqrt{3}}$

Solve each equation.

20. $\sqrt{8r + 4} = 3\sqrt{r}$

21. $2\sqrt{z} = z + 1$

22. $\sqrt{3x - 2} = x - 4$

Simplify each expression. Write answers with only positive exponents.

23. $16^{3/4}$

24. $11^{2/3} \cdot 11^{4/3}$

25. $\dfrac{(5^{1/2})^3}{5^{-1/2}}$

13. _____

14. _____

15. _____

16. _____

17. _____

18. _____

19. _____

20. _____

21. _____

22. _____

23. _____

24. _____

25. _____

CHAPTER 9 QUADRATIC EQUATIONS

9.1 Solving Quadratic Equations by the Square Root Property

1️⃣ Solve equations of the form $x^2 =$ a number. (See Frames 1–6 below.)

2️⃣ Solve equations of the form $(ax + b)^2 =$ a number. (Frames 7–21)

Before you begin this section, review Section 4.5, "Solving Quadratic Equations by Factoring" in your textbook and in your Study Guide.

---

1. An equation containing a squared term, such as the equation $x^2 + 2x - 24 = 0$, is called a _____ equation. Then $3x^2 + 2x + 1 = 0$, $4x^2 = 9$, and $3x^2 = -5x$ are all examples of _____ equations. | quadratic

quadratic

2. You already have used one method of solving quadratic equations (in Section 4.5), which is the method of _____. To solve the quadratic equation $x^2 + 2x - 24 = 0$, factor it.

   $(\quad\quad)(\quad\quad) = 0$

   By the zero-factor property,

   _____ $= 0$ or _____ $= 0$,

   from which

   $x =$ ____ or $x =$ ____ .

   Most quadratic equations have ____ different solutions. | factoring

x + 6; x − 4

−6; 4

two

3. Some quadratic equations can be solved by taking square roots of _____ sides of the equation. This method is justified by the square root _____ of equations, which says that if b is a _____ number and if $a^2 = b$, then $a = \sqrt{b}$ and $a =$ _____. For example, to solve the equation

   $k^2 = 5$, | both

property

positive

$-\sqrt{b}$

9.1 Solving Quadratic Equations by the Square Root Property

    take the _____ _____ of both sides. | square roots
$$k = \_\_\_\_ \quad \text{or} \quad k = \_\_\_\_$$ | $\sqrt{5}$; $-\sqrt{5}$

Solve each equation in Frames 4–6.

4. $z^2 = 100$

$$z = \_\_\_\_ \quad \text{or} \quad z = \_\_\_\_$$ | 10; −10

5. $p^2 = 72$

$$p = \_\_\_\_ \quad \text{or} \quad p = \_\_\_\_$$ | $\sqrt{72}$; $-\sqrt{72}$

    Simplify to get $p = \_\_\_\_$ or $p = \_\_\_\_$. | $6\sqrt{2}$; $-6\sqrt{2}$

6. $y^2 = -9$

    _____ | No real solutions

7. To solve the equation

$$(x - 5)^2 = 36,$$

    use the _____ _____ property. This leads to two possibilities: | square root

$$x - 5 = \_\_\_\_ \quad \text{or} \quad x - 5 = \_\_\_\_.$$ | 6; −6

    Solve these two equations:

$$x = \_\_\_\_ \quad \text{or} \quad x = \_\_\_\_.$$ | 11; −1

Solve each equation in Frames 8–20 using the square root property.

8. $(x + 3)^2 = 25$

$$\_\_\_\_ = \_\_\_\_ \quad \text{or} \quad x + 3 = \_\_\_\_$$ | x + 3; 5; −5
$$x = \_\_\_\_ \quad \text{or} \quad x = \_\_\_\_$$ | 2; −8

9. $(m - 8)^2 = 81$

$$m = \_\_\_\_ \quad \text{or} \quad m = \_\_\_\_$$ | 17; −1

10. $(2x + 1)^2 = 36$

        2x + 1 = _____   or   2x + 1 = _____   6; −6

             2x = _____   or   2x = _____   5; −7

              x = _____   or   x = _____   5/2; −7/2

11. $(3r - 5)^2 = 121$

        3r − 5 = _____   or   3r − 5 = _____   11; −11

             3r = _____   or   3r = _____   16; −6

              r = _____   or   r = _____   16/3; −2

12. $(2z - 9)^2 = 16$

              z = _____   or   z = _____   13/2; 5/2

13. $(x - 8)^2 = 15$

          x − 8 = _____   or   x − 8 = _____ $\sqrt{15}$; $-\sqrt{15}$

              x = _____   or   x = _____ $8 + \sqrt{15}$; $8 - \sqrt{15}$

14. $(q + 3)^2 = 7$

          q + 3 = _____   or   q + 3 = _____ $\sqrt{7}$; $-\sqrt{7}$

              q = _____   or   q = _____ $-3 + \sqrt{7}$; $-3 - \sqrt{7}$

15. $(r - 6)^2 = 21$

              r = _____   or   r = _____ $6 + \sqrt{21}$; $6 - \sqrt{21}$

16. $(x - 4)^2 = 12$

          x − 4 = _____   or   x − 4 = _____ $\sqrt{12}$; $-\sqrt{12}$

We can simplify $\sqrt{12}$ as $\sqrt{12} = \sqrt{4(\quad)} =$ _____.   3; $2\sqrt{3}$
Therefore,

          x − 4 = _____   or   x − 4 = _____ $2\sqrt{3}$; $-2\sqrt{3}$

              x = _____   or   x = _____ . $4 + 2\sqrt{3}$; $4 - 2\sqrt{3}$

9.2 Solving Quadratic Equations by Completing the Square

17. $(x + 5)^2 = 32$

    $x + 5 =$ _____ or $x + 5 =$ _____ | $\sqrt{32}$; $-\sqrt{32}$

    $x + 5 =$ _____ or $x + 5 =$ _____ | $4\sqrt{2}$; $-4\sqrt{2}$

    $x =$ _____ or $x =$ _____ | $-5 + 4\sqrt{2}$; $-5 - 4\sqrt{2}$

18. $(p - 3)^2 = 24$

    $p =$ _____ or $p =$ _____ | $3 + 2\sqrt{6}$; $3 - 2\sqrt{6}$

19. $(2x - 5)^2 = 18$

    $2x - 5 =$ _____ or $2x - 5 =$ _____ | $\sqrt{18}$; $-\sqrt{18}$

    $2x - 5 =$ _____ or $2x - 5 =$ _____ | $3\sqrt{2}$; $-3\sqrt{2}$

    $2x =$ _____ or $2x =$ _____ | $5 + 3\sqrt{2}$; $5 - 3\sqrt{2}$

    $x =$ _____ or $x =$ _____ | $\dfrac{5 + 3\sqrt{2}}{2}$; $\dfrac{5 - 3\sqrt{2}}{2}$

20. $(5y - 3)^2 = 72$

    $y =$ _____ or $y =$ _____ | $\dfrac{3 + 6\sqrt{2}}{5}$; $\dfrac{3 - 6\sqrt{2}}{5}$

21. $(3x + 7)^2 = -14$

    The square root of -14 (*is/is not*) a real number. Therefore, there is (*one/no*) real number solution for this equation. | is not; no

---

9.2 Solving Quadratic Equations by Completing the Square

[1] Solve quadratic equations by completing the square when the coefficient of the squared term is 1. (See Frames 1-6 below.)

[2] Solve quadratic equations by completing the square when the coefficient of the squared term is not 1. (Frames 7-8)

[3] Simplify an equation before solving. (Frame 9)

Chapter 9 Quadratic Equations

1. The method of completing the square lets you begin with any quadratic equation, and end with a form such as $(x - 1)^2 = 5$, which can be solved by using the square root property. For example, solve the quadratic equation

$$x^2 - 2x - 4 = 0.$$

You want only terms with variables on the left side. First add _____ to both sides. The resulting equation is | 4

$$\underline{\hspace{2cm}} = \underline{\hspace{1cm}}.$$ | $x^2 - 2x$; 4

Then "complete the square." To do this, take half the coefficient of x and square it. The coefficient of x is _____; half the coefficient is _____; | -2; -1
the square of half the coefficient is _____. Add | 1
_____ on both sides of the equation that you got above. | 1

$$x^2 - 2x + \underline{\hspace{1cm}} = 4 + \underline{\hspace{1cm}}$$ | 1; 1
$$\underline{\hspace{3cm}} = \underline{\hspace{1cm}}$$ | $x^2 - 2x + 1$; 5

The left-hand side is a perfect _____ trinomial. Write it in factored form. | square

$$x^2 - 2x + 1 = (\quad\quad)(\quad\quad)$$ | $x - 1$; $x - 1$
$$= (\quad\quad)^2$$ | $x - 1$

This process is called _____ the square. | completing
Now you can solve the quadratic equation $(\quad)^2 =$ | $x - 1$
_____ by using the square root property as was done in the previous section. | 5

$$x - 1 = \underline{\hspace{1cm}} \quad \text{or} \quad x - 1 = \underline{\hspace{1cm}}$$ | $\sqrt{5}$; $-\sqrt{5}$
$$x = \underline{\hspace{1cm}} \quad \text{or} \quad x = \underline{\hspace{1cm}}$$ | $1 + \sqrt{5}$; $1 - \sqrt{5}$

9.2 Solving Quadratic Equations by Completing the Square

Solve the quadratic equations in Frames 2–8 by completing the square.

2. $r^2 + 6r = -2$

   Half of _____ is _____. Square 3, getting _____. | 6; 3; 9
   Add _____ on both sides. | 9

   $r^2 + 6r +$ _____ $= -2 +$ _____. | 9; 9

   Factor the left side as a perfect square.

   (_____)$^2 =$ _____ | r + 3; 7

   Solve this equation.

   r = _____ or r = _____ | $-3 + \sqrt{7}$; $-3 - \sqrt{7}$

3. $x^2 - 8x = -5$

   Half of _____ is _____. Add _____ to both sides. | -8; -4; $(-4)^2 = 16$

   $x^2 - 8x +$ _____ $= -5 +$ _____ | 16; 16
   (_____)$^2 =$ _____ | x − 4; 11

   x = _____ or x = _____ | $4 + \sqrt{11}$; $4 - \sqrt{11}$

4. $q^2 + 6q + 1 = 0$

   Rewrite the equation as $q^2 + 6q =$ _____. Solve the equation. | -1

   q = _____ or q = _____ | $-3 + 2\sqrt{2}$; $-3 - 2\sqrt{2}$

5. $x^2 - 10x - 20 = 0$

   x = _____ or x = _____ | $5 + 3\sqrt{5}$; $5 - 3\sqrt{5}$

6. $z^2 - 3z = 10$

   Half of _____ is _____. Add _____ to both sides. | -3; $-3/2$; $9/4$

   $z^2 - 3z +$ _____ $= 10 +$ _____ | 9/4; 9/4
   (_____)$^2 =$ _____ | z − 3/2; 49/4

   z = _____ or z = _____ | 5; −2

7. $2x^2 - 5x = 5$

   The coefficient of the squared term is not 1. In this case you should divide through by ____. | 2

   $$x^2 - \frac{5}{2}x = \underline{}$$ | 5/2

   Proceed as before. Half of ____ is ____. | -5/2; -5/4

   $$x^2 - \frac{5}{2}x + \underline{} = \frac{5}{2} + \underline{}$$
   $$()^2 = \underline{}$$ | $\frac{25}{16}$; $\frac{25}{16}$
   x - 5/4; 65/16

   $x = \underline{}$ or $x = \underline{}$ | $\frac{5 + \sqrt{65}}{4}$; $\frac{5 - \sqrt{65}}{4}$

8. $3m^2 - 9m = -1$

   $m = \underline{}$ or $m = \underline{}$ | $\frac{3}{2} + \frac{\sqrt{69}}{6}$; $\frac{3}{2} - \frac{\sqrt{69}}{6}$

9. $(x - 6)(x + 2) = 9$

   This equation must be _____ before you are able to solve it. | simplified

   Multiply using FOIL.

   _____ = 9 | $x^2 - 4x - 12$

   To get only variable terms on the left side, add ____ to both sides. | 12

   _____ = _____ | $x^2 - 4x$; 21

   To get a perfect square on the left, add ____ to both sides. | 4

   _____ = _____ | $x^2 - 4x + 4$; 21 + 4

   Proceed as before to get the solution.

   $x = \underline{}$ or $x = \underline{}$ | 7; -3

9.3 Solving Quadratic Equations by the Quadratic Formula

[1] Identify the values of a, b, and c in quadratic equation. (See Frames 1-6 below.)

[2] Use the quadratic formula to solve quadratic equations. (Frames 7-17)

[3] Solve quadratic equations with only one solution. (Frame 18)

[4] Solve quadratic equations with fractions. (Frames 19-20)

[5] Use the quadratic equation to solve an applied problem. (Frame 21)

---

1. A quadratic equation is in _____ form if it is in the form

    $$ax^2 + bx + c = 0.$$

   standard

2. The equation

    $$x^2 - 9x + 20 = 0$$

   (*is/is not*) in standard form. The values of a, b, and c are

    a = ____, b = ____, and c = ____.

   Note that the coefficient of x^2 is understood to be ____.

   is

   1; −9; 20

   1

Find the values of a, b, and c in Frames 3-6.

3. $m^2 - 4m - 5$

   Rewrite the equation with ____ on one side.

    $$\text{_____} = 0$$

    a = ____, b = ____, c = ____

   0

   $m^2 - 4m + 5$

   1; −4; 5

4. $8p^2 + 5p = -6$

    a = ____, b = ____, c = ____

   8; 5; 6

5. $12y^2 - 5y = 0$

   $a =$ _____, $b =$ _____, $c =$ _____ | 12; −5; 0

6. $m^2 = 16$

   $a =$ _____, $b =$ _____, $c =$ _____ | 1; 0; −16

7. The solutions of a quadratic equation in standard form

   $$ax^2 + bx + c = 0,$$

   where a, b, and c are _____ numbers and $a \neq$ _____, | real; 0
   are given by

   $$x = \frac{\pm \sqrt{}}{2a}.$$ | −b; b² − 4ac

   This formula is called the _____ formula. You should memorize it. | quadratic

8. To solve the equation $x^2 - 2x - 15 = 0$, first identify the values of a, b, and c. Here a is the coefficient of the _____ term, or $a =$ _____. | x²; 1
   Also, $b =$ _____ and $c =$ _____. Substitute these values into the quadratic formula. | −2; −15

   $$x = \frac{-() \pm \sqrt{()^2 - 4()()}}{2()}$$ | −2; −2; 1; −15; 1

   $$= \frac{\pm \sqrt{ + }}{2}$$ | 2; 4; 60; 2

   $$= \frac{2 \pm \sqrt{}}{2}$$ | 64

   $$= \frac{2 \pm }{2}$$ | 8

   This result leads to _____ different solutions. With the + sign, the solution is | two

   $$x = \frac{2 + }{2} = \frac{}{2} = \underline{}.$$ | 8; 10; 5

9.3 Solving Quadratic Equations by the Quadratic Formula

With the − sign, the solution is

$$x = \frac{2 - \underline{}}{2} = \underline{}$$ 8; −3

The solutions of the original equation are

$$x = \underline{} \quad \text{or} \quad x = \underline{}.$$ 5; −3

Solve the equations in Frames 8–15.

9. $2m^2 + 7m - 4 = 0$

   Here $a = \underline{}$, $b = \underline{}$, and $c = \underline{}$. By the quadratic formula, 2; 7; −4

   $$m = \frac{-() \pm \sqrt{()^2 - 4()()}}{2()}$$ 7; 7; 2; −4 ; 2

   $$= \frac{\underline{} \pm \sqrt{\underline{} + \underline{}}}{4}$$ −7; 49; 32

   $$= \frac{\underline{} \pm \sqrt{\underline{}}}{4}$$ −7; 81

   $$= \frac{-7 \pm \underline{}}{4}.$$ 9

   With the + sign, $m = \dfrac{-7 + \underline{}}{4} = \underline{}$. 9; 1/2

   With the − sign, $m = \dfrac{-7 - \underline{}}{4} = \underline{}$. 9; −4

10. $3x^2 + x - 2 = 0$

    $a = \underline{}$, $b = \underline{}$, and $c = \underline{}$. 3; 1; −2

    $$x = \frac{-() \pm \sqrt{()^2 - 4()()}}{2()}$$ 1; 1; 3; −2 ; 3

    $$x = \frac{\underline{} \pm \sqrt{\underline{}}}{6}$$ −1; 25

    $$= \frac{-1 \pm \underline{}}{6}$$ 5

    From this result, we find that

    $$x = \underline{} \quad \text{or} \quad x = \underline{}.$$ 2/3; −1

11. $x^2 - 2x = 4$

> Rewrite the equation so that one side is ____. | 0
>
> $$x^2 - 2x\ \underline{\quad} = 0$$ | -4
>
> Here $a =$ ____, $b =$ ____, $c =$ ____. | $1;\ -2;\ -4$
>
> $$x = \frac{-(\quad) \pm \sqrt{(\quad)^2 - 4(\quad)(\quad)}}{2(\quad)}$$ | $-2;\ -2;\ 1;\ -4$
>
> $$= \frac{\underline{\quad} \pm \sqrt{\underline{\quad}}}{2}$$ | $2;\ 20$
>
> Since $\sqrt{20} = \sqrt{4(\quad)} = \underline{\quad}$, | $5;\ 2\sqrt{5}$
>
> $$x = \frac{2 \pm \underline{\quad}}{2}.$$ | $2\sqrt{5}$
>
> The common factor of the numerator is ____. Factor out the common factor. | 2
>
> $$x = \frac{2(\underline{\quad})}{2} = \underline{\quad}$$ | $1 \pm \sqrt{5};\ 1 \pm \sqrt{5}$
>
> The two solutions here are ____ or ____. | $1 + \sqrt{5};\ 1 - \sqrt{5}$

12. $q^2 + 1 = 6q$

> Here $a =$ ____, $b =$ ____, and $c =$ ____. | $1;\ -6;\ 1$
>
> $$x = \frac{6 \pm \sqrt{(-6)^2 - \underline{\quad}}}{2}$$ | 4
>
> $$= \frac{6 \pm \sqrt{\underline{\quad}}}{2}$$ | 32
>
> $$= \frac{6 \pm \underline{\quad}\sqrt{\underline{\quad}}}{2}$$ | $4;\ 2$
>
> $$= \frac{(\underline{\quad})}{2}$$ | $2;\ 3 \pm 2\sqrt{2}$
>
> $$= \underline{\quad}$$ | $3 \pm 2\sqrt{2}$

9.3 Solving Quadratic Equations by the Quadratic Formula

13. $4x^2 - 4x - 11 = 0$

    $a = \underline{}$, $b = \underline{}$, $c = \underline{}$ 4; −4; −11

    $x = \dfrac{-() \pm \sqrt{()^2 - 4()()}}{8}$ −4; −4; 4; −11

    $= \dfrac{4 \pm \sqrt{}}{8}$ 192

    $= \dfrac{4 \pm \sqrt{3}}{8}$ 8

    $= \dfrac{()}{8}$ 4; $1 \pm 2\sqrt{3}$

    $= \underline{}$ $\dfrac{1 \pm 2\sqrt{3}}{2}$

14. $4z^2 + 12z + 1 = 0$

    $z = \underline{}$ $\dfrac{-3 \pm 2\sqrt{2}}{2}$

15. $2x^2 + 2x - 3 = 0$

    $x = \underline{}$ $\dfrac{-1 \pm \sqrt{7}}{2}$

16. $x^2 - 4x + 5 = 0$

    $a = \underline{}$, $b = \underline{}$, $c = \underline{}$ 1; −4; 5

    $x = \dfrac{-() \pm \sqrt{()^2 - 4()()}}{2}$ −4; −4; 1; 5

    $= \dfrac{4 \pm \sqrt{}}{2}$ −4

    $\sqrt{-4}$ (*is/is not*) a real number. (The square   is not
    roots of negative numbers will be discussed in
    Section 9.4, which follows this section.)

17. Does $y^2 - 6y + 14 = 0$ have any real number solutions? By the quadratic formula,

    $y = \dfrac{6 \pm \sqrt{()^2 - 4()()}}{2}$ −6; 1; 14

    $= \dfrac{6 \pm \sqrt{}}{2}$.   −20

$\sqrt{-20}$ (is/is not) a real number, so that the equation has no _____ solutions.

	is not
	real

18. Solve $9x^2 + 4 = 12x$.

    Rewrite the equation in standard form.

    $$\underline{} = 0$$

    Here, a = ____, b = ____, and c = ____.

    $$x = \frac{-() \pm \sqrt{()^2 - 4()()}}{2()}$$

    $$= \frac{ \pm \sqrt{}}{18}$$

    $$= \underline{}$$

    How many solutions are there? _____
    The trinomial _____ is a perfect square.

	$9x^2 - 12x + 4$
	9; −12; 4
	−12; −12; 9; 4
	9
	12; 0
	$\frac{2}{3}$
	1
	$9x^2 - 12x + 4$

19. Solve $\frac{1}{2}m^2 = m + 2$.

    Multiply both sides by ____

    $$\underline{} = \underline{}$$

    Rewrite the equation so that one side equals ____.

    $$\underline{} = 0$$

    Solve the equation.

    $$m = \underline{} \quad \text{or} \quad m = \underline{}$$

	2
	m^2; $2m + 4$
	0
	$m^2 - 2m - 4$
	$1 + \sqrt{5}$; $1 - \sqrt{5}$

20. Solve $\frac{2}{3}z^2 - \frac{1}{9}z = 1$.

    Here, multiply by the ____ of the fractions.

    $$z = \underline{} \quad \text{or} \quad z = \underline{}$$

	LCD
	$\frac{1 + \sqrt{217}}{12}$; $\frac{1 - \sqrt{217}}{12}$

9.3 Solving Quadratic Equations by the Quadratic Formula

21. It takes Marcie 1 hour longer than Ron to wash the walls of a certain room. If it takes them 1 1/5 hours, or 6/5 hours, to do the job together, how long would it take Marcie working alone?

    Let x = the time it would take Marcie working alone.

    Then _____ = the time it would take Ron working alone. $x - 1$

    Make a chart.

	Rate	Time working together	Fractional part done when working together
Marcie			
Ron			

Rate	Time	Fractional part
$\frac{1}{x}$	$\frac{6}{5}$	$\frac{6/5}{x}$
$\frac{1}{x-1}$	$\frac{6}{5}$	$\frac{6/5}{x-1}$

    Write an equation.

    _____ + _____ = 1 $\frac{6/5}{x}$; $\frac{6/5}{x-1}$

    Simplify the fractions on the left side.

    $\frac{6}{\text{\_\_\_}} + \frac{6}{\text{\_\_\_}} = 1$ $5x$; $5(x-1)$

    The LCD is ___ · ___ · _____ 5; x; $(x-1)$

    Multiply both sides of the equation by the LCD.

    _____ + _____ = _____ $6(x-1)$; $6x$; $5(x)(x-1)$

    Write the equation in standard form.

    _____ − _____ + _____ = 0 $5x^2$; $17x$; 6

294   Chapter 9   Quadratic Equations

Solve using the quadratic formula.	
a = ____, b = ____, c = ____	5; −17; 6
$$x = \frac{-(\quad) \pm \sqrt{(\quad)^2 - 4(\quad)(\quad)}}{2(\quad)}$$	−17; −17; 5; 6 5
$$x = \frac{\pm \sqrt{\quad}}{\quad}$$	17; 169 10
$$= \frac{\pm}{\quad}$$	17; 13 10
= ____ or x = ____	3; 2/5
Discard ____ because this solution would give a negative time for Ron, which is impossible.	2/5
It would take Marcie ___ hours working alone to complete the job.	3

9.4 Complex Numbers

[1] Write complex numbers like $\sqrt{-5}$ as multiples of i. (See Frames 1–16 below.)

[2] Add and subtract complex numbers. (Frames 17–19)

[3] Multiply complex numbers. (Frames 20–23)

[4] Write complex number quotients in standard form. (Frames 24–25)

[5] Solve quadratic equations with complex number solutions. (Frames 26–35)

1. In Frames 16 and 17 of the previous section, the square roots of negative numbers turned up in the solutions of quadratic equations by the quadratic formula. The quadratic equations in Frames 16 and 17 have no _____ number solutions. Does this mean that these equations have no solution at all?	real

9.4 Complex Numbers   295

2. To answer the question, you have to learn about a set of numbers beyond the real numbers, the complex numbers. First of all, define the number i as

$$i = \underline{}.$$

$\sqrt{-1}$

The number i is called an _____ number. It (*is/is not*) a real number, since there is no real number whose square is _____.

imaginary
is not
-1

3. The square roots of negative numbers turn out to be multiples of i, by the following rule:

$$\sqrt{-a} = \sqrt{a \cdot (-1)} = \sqrt{a} \cdot \sqrt{-1} = \underline{}.$$

$\sqrt{a} \cdot i$ or $i\sqrt{a}$

(Make sure you write the number i outside the radical.) For example,

$\sqrt{-36} = \sqrt{} \cdot i = \underline{};$ 36; 6i

$\sqrt{-4} = \sqrt{} \cdot i = \underline{};$ 4; 2i

$\sqrt{-30} = \sqrt{} \cdot i = \underline{};$ 30; $i\sqrt{30}$

$\sqrt{-18} = \sqrt{} \cdot i = \underline{};$ 18; $3i\sqrt{2}$

$3\sqrt{-50} = \underline{}.$ $3i\sqrt{50} = 15i\sqrt{2}$

Find the square roots in Frames 4-10.

4. $\sqrt{-49} = \sqrt{} \cdot i = \underline{}$ 49; 7i

5. $\sqrt{-81} = \underline{}$ 9i

6. $\sqrt{-121} = \underline{}$ 11i

7. $\sqrt{-12} = \sqrt{} \cdot i = \sqrt{4 \cdot } \cdot i = \underline{}$ 12; 3; $2i\sqrt{3}$

8. $\sqrt{-48} = \sqrt{48} \cdot i = \underline{}$ $4i\sqrt{3}$

296   Chapter 9   Quadratic Equations

9. $\sqrt{-75}$ = _____ $5i\sqrt{3}$

10. $\sqrt{-98}$ = _____ $7i\sqrt{2}$

Simplify the expressions in Frames 11 and 12.

11. $4\sqrt{-48} + 2\sqrt{-27}$ = 4( ) + 2( ) $4i\sqrt{3}$; $3i\sqrt{3}$

   = _____ + _____ $16i\sqrt{3}$; $6i\sqrt{3}$

   = _____ $22i\sqrt{3}$

12. $-2\sqrt{-18} + 3\sqrt{-50} + 4\sqrt{-72}$

   = $-2($ $)+ 3($ $) + 4($ $)$ $3i\sqrt{2}$; $5i\sqrt{2}$; $6i\sqrt{2}$

   = _____ + _____ + _____ $-6i\sqrt{2}$; $15i\sqrt{2}$; $24i\sqrt{2}$

   = _____ $33i\sqrt{2}$

13. The sum of a real number and a real number multiple of i is called a _____ number. For example, 8 + 2i, 6 − 3i, and −4 − 5i are all examples of _____ numbers. But real numbers such as 3, $\sqrt{7}$, 0, and −8 · 1 are also _____ numbers since they can be written in the form a + bi, with b = 0. If b ≠ 0, then a + bi is also called a(n) _____ number. A complex number written in the form a + bi is said to be in _____ form.

   complex

   complex

   complex

   imaginary
   standard

14. 12 + 0i, which equals _____, is a complex number. In fact, every _____ number is a _____ number.

   12
   real; complex

15. 0 − 5i, which equals _____, is a complex number. 0 − 5i also is a(n) _____ number.

   −5i
   imaginary

9.4 Complex Numbers   297

16. $-7 + 6i$ is a _____ number. | complex
    $-7 + 6i$ also is a(n) _____ number. | imaginary

17. In the number $-7 + 6i$, -7 is called the _____ | real
    part and _____ is called the imaginary part. | $6i$

18. To add $4 - 2i$ and $5 + 7i$, add the real parts _____ | 4
    and ___ and the imaginary parts _____ and _____. | 5; $-2i$; $7i$

    $(4 - 2i) + (5 + 7i)$
    $= ($ $) + ($ $)$ | $4 + 5$; $-2i + 7i$
    $= $ _____ | $9 + 5i$

19. To subtract $4 - 5i$ from $6 - 2i$, change $4 - 5i$ to
    its _____ and add. | negative

    $(6 - 2i) - (4 - 5i) = (6 - 2i) + ($ $)$ | $-4 + 5i$
    $= $ _____ | $2 + 3i$

20. The product of $2 - i$ and $3 + 4i$ is

    $(2 - i)(3 + 4i)$
    $= 2 \cdot 3 + 2($ $) + (-i)(3) + (-i)($ $)$ FOIL | $4i$; $4i$
    $= 6 + $ _____ $- 3i - $ _____ . | $8i$; $4i^2$

    Since $i^2 = $ _____ , we get | -1

    $= 6 + 8i - 3i + $ _____ | 4
    $= $ _____ . | $10 + 5i$

Find each product in Frames 21–23. Write the answers
in standard form.

21. $(2 - 3i)(1 - i) = $ _____ | $-1 - 5i$

22. $(5 - i)(6 + 3i) = $ _____ | $33 + 9i$

23. $(6 - 5i)(6 + 5i) = $ _____ | 61

24. The conjugate of the complex number a + bi is
_____. Thus, the _____ of 1 + i is
1 − i. | a − bi; conjugate

To find the quotient

$$\frac{2-i}{1+i},$$

multiply the numerator and the denominator by the
_____ of the denominator, _____. | conjugate; 1 − i

$$\frac{(2-i)(1-i)}{(1+i)(1-i)} = \underline{}$$ | $\frac{1-3i}{2}$

In standard form, this number is written

_____. | $\frac{1}{2} - \frac{3}{2}i$

25. Find $\frac{3+2i}{1-4i}$. Write the answer in standard form.

_____. | $-\frac{5}{17} + \frac{14}{17}i$

26. You already used the quadratic formula to find
the solution of certain quadratic equations.
If $ax^2 + bx + c = 0$ (where $a \neq 0$), then

$$x = \frac{\underline{} \pm \sqrt{\underline{}}}{2a}$$ | −b; $b^2 − 4ac$

27. The quadratic formula can also be used to find
imaginary number solutions for quadratic equations.
For example, to find the solutions of

$$x^2 - 4x + 5 = 0,$$

we let a = ____, b = ____, and c = ____. Using
the quadratic formula gives | 1; −4; 5

$$x = \frac{-() \pm \sqrt{()^2 - 4()()}}{2()}$$ | −4; −4; 1; 5
1

$$= \frac{4 \pm \sqrt{}}{2}.$$ | −4

As discussed above, $\sqrt{-4} = \sqrt{} \cdot i = \underline{}$. | 4; 2i

9.4 Complex Numbers

Therefore,

$$x = \frac{4 \pm \underline{}}{2}.$$ | $2i$

Factor out ___. | 2

$$x = \frac{()}{2} = \underline{}$$ | 2; $2 \pm i$; $2 \pm i$

Solve the equations in Frames 28–35. Write solutions in standard form.

28. $p^2 + 6p + 25 = 0$

$$p = \frac{-() \pm \sqrt{()^2 - 4()()}}{2()}$$ | 6; 6; 1; 25
1

$$= \frac{-6 \pm \sqrt{}}{2}$$ | -64

$$= \frac{-6 \pm }{2}$$ | $8i$

$$= \underline{}$$ | $-3 \pm 4i$

29. $z^2 + 4z + 20 = 0$

$$z = \frac{-() \pm \sqrt{()^2 - 4()()}}{2()}$$ | 4; 4; 1; 20
1

$$= \frac{-4 \pm \sqrt{}}{2}$$ | -64

$$= \frac{-4 \pm }{2}$$ | $8i$

$$= \underline{}$$ | $-2 \pm 4i$

30. $x^2 - 6x + 12 = 0$

$$x = \frac{-() \pm \sqrt{()^2 - 4()()}}{2()}$$ | -6; -6; 1; 12
1

$$= \frac{ \pm \sqrt{}}{2}$$ | 6; -12

$$= \frac{6 \pm }{2}$$ | $2i\sqrt{3}$

$$= \underline{}$$ | $3 \pm i\sqrt{3}$

Chapter 9 Quadratic Equations

31. $q^2 + 10q + 27 = 0$

$$q = \frac{\_\_\_ \pm \sqrt{\_\_\_}}{2}$$ 　　　$-10;\ -8$

$$= \frac{-10 \pm \sqrt{\_\_\_}}{2}$$ 　　　$2i\sqrt{2}$

$$= _____$$ 　　　$-5 \pm i\sqrt{2}$

32. $4x^2 - 4x + 5 = 0$

$$x = \frac{-(\ ) \pm \sqrt{(\ )^2 - 4(\ )(\ )}}{2(\ )}$$ 　　　$-4;\ -4;\ 4;\ 5$; 4

$$= \frac{\_\_\_ \pm \sqrt{\_\_\_}}{8}$$ 　　　$4;\ -64$

$$= \frac{4 \pm \_\_\_}{8}$$ 　　　$8i$

$$= _____$$ 　　　$\frac{1 \pm 2i}{2} = \frac{1}{2} \pm i$

33. $2r^2 + 6r + 9 = 0$

$$r = \frac{\_\_\_ \pm \sqrt{\_\_\_}}{4}$$ 　　　$-6;\ -36$

$$= \frac{\_\_\_ \pm \_\_\_}{4}$$ 　　　$-6;\ 6i$

$$= _____$$ 　　　$\frac{-3 \pm 3i}{2} = -\frac{3}{2} \pm \frac{3}{2}i$

34. $3x^2 - 2x + 3 = 0$

$$x = \frac{\_\_\_ \pm \sqrt{\_\_\_}}{6}$$ 　　　$2;\ -32$

$$= \frac{\_\_\_ \pm \_\_\_}{6}$$ 　　　$2;\ 4i\sqrt{2}$

$$= _____$$ 　　　$\frac{1}{3} \pm \frac{2\sqrt{2}}{3}i$

35. $4m^2 - 12m + 21 = 0$

$$m = _____$$ 　　　$\frac{3 \pm 2i\sqrt{3}}{2} = \frac{3}{2} \pm i\sqrt{3}$

9.5 Graphing Quadratic Equations in Two Variables

[1] Graph quadratic equtions of the form $y = ax^2 + bx + c$ ($a \neq 0$). (See Frames 1-5 below.)

[2] Find the vertex of a parabola. (Frames 6-10)

[3] Use a graph to determine the number of real solutions of a quadratic equation. (Frames 11-17)

---

1. To graph $y = x^2$, select a number of values for x, and find the corresponding values for y. Complete the following table and write the ordered pairs.

x	y	Ordered pair
-3	9	(-3, 9)
-2	___	_____
-1	___	_____
0	___	_____
1	___	_____
2	___	_____
3	___	_____

4; (-2, 4)
1; (-1, 1)
0; (0, 0)
1; (1, 1)
4; (2, 4)
9; (3, 9)

Graph these ordered pairs on the coordinate system shown here, and draw a smooth curve through them.

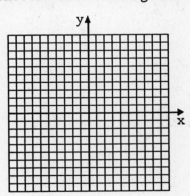
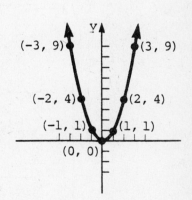

This graph is called a _____. The lowest point on the graph is at the _____. An equation of the form $y = ax^2 + bx + c$ is called a _____ _____. Is $y = x^2$ a quadratic function? _____

parabola
origin

quadratic; function
yes

302 Chapter 9 Quadratic Equations

2. To graph the parabola $y = x^2 + 4$, note that the term 4 causes the graph to be _____ units above the graph in Frame 1. Complete the table of values at the left, and graph ordered pairs to get the parabola for $y = x^2 + 4$.

x	y
-2	___
-1	___
0	___
1	___
2	___

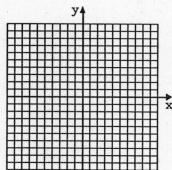

4
8
5
4
5
8

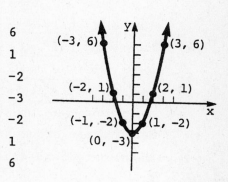

3. Graph $y = x^2 - 3$

x	y
-3	___
-2	___
-1	___
0	___
1	___
2	___
3	___

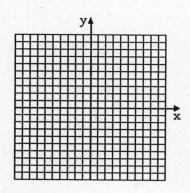

6
1
-2
-3
-2
1
6

4. Graph $y = -x^2 + 5$.

x	y
-3	___
-2	___
-1	___
0	___
1	___
2	___
3	___

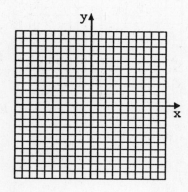

-4
1
4
5
4
1
-4

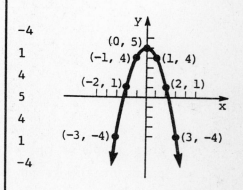

9.5 Graphing Quadratic Equations in Two Variables

5. Graph $y = -x^2 - 2$.

x	y
-3	____
-2	____
-1	____
0	____
1	____
2	____
3	____

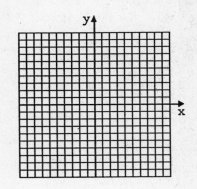

-11
-6
-3
-2
-3
-6
-11

(0, -2)
(-1, -3) (1, -3)
(-2, -6) (2, -6)
(-3, -11) (3, -11)

Notice that you must extend the grid at the bottom in order to graph the ordered pairs _____ and _____.

(-3, -11)
(3, -11)

6. The most important single point to find in graphing a quadratic equation is the _____ of the parabola. Notice in Frames 3 and 4, the vertex is exactly halfway between the _____. If the graph of a quadratic equation $y = ax^2 + bx + c$ has two x-intercepts, we should be able to find the vertex. But the x-intercepts are found by letting $y =$ ____. Therefore, the x-values of the x-intercepts are the _____ of the equation $0 = ax^2 + bx + c$.

vertex

x-intercepts.

0

solutions

According to the quadratic formula, these are

$x =$ _____ and

$x =$ _____.

$$\frac{-b + \sqrt{b^2 - 4ac}}{2a}$$

$$\frac{-b - \sqrt{b^2 - 4ac}}{2a}$$

The value of x that is halfway between these numbers is

$$x = \frac{1}{2}\left(\frac{-b + \sqrt{b^2 - 4ac}}{2a} + \frac{-b - \sqrt{b^2 - 4ac}}{2a}\right)$$

$$= \frac{1}{2}\left(\frac{-b + \sqrt{b^2 - 4ac} - \underline{} - \underline{}}{2a}\right)$$

$$= \frac{1}{2}\left(\frac{\underline{}}{2a}\right) = \underline{}$$

b; $\sqrt{b^2 - 4ac}$

$-2b$; $-\dfrac{b}{2a}$

304   Chapter 9   Quadratic Equations

This is the x-value of the _____ of the graph of $y = ax^2 + bx + c$. To find the y-value of the vertex, substitute the value for _____ in the equation and solve for y.

vertex

$-\dfrac{b}{2a}$

7. After the vertex and the x-intercepts, another important point to be found in graphing a quadratic equation is the _____. Find the y-value of the _____ by letting x = _____ in the equation and solving for y. Finally, find and plot additional _____ _____, as needed, to draw a smooth curve.

y-intercept
y-intercept
0
ordered
pairs

8. Identify the vertex and the intercepts of $y = x^2 - 2x - 2$. Then graph this equation.

If $y = 0$, the equation is _____.

$0 = x^2 - 2x - 2$

To find the x-value of the vertex, use x = _____, with a = _____ and b = _____. The x-value of the vertex is $-\dfrac{}{2(1)}$, or _____. To find the y-value of the vertex, let x = _____ in the equation and solve for _____. The vertex is the point _____.

$-\dfrac{b}{2a}$;
1; -2
-2; 1
1
y; (1, -3)

The x-values of the x-intercepts are the _____ of $0 = x^2 - 2x - 2$, or

solutions

x = _____ and x = _____, or,

$1 + \sqrt{3}$; $1 - \sqrt{3}$

to the nearest tenth, _____ and _____.

2.7; -.7

The x-intercepts are the ordered pairs, _____ and _____.

(2.7, 0)
(-.7, 0)

To find the y-value of the y-intercept, solve the following equation for _____.

y

$y = (\ \ )^2 - 2(\ \ ) - 2$
$y = $ _____

0; 0
-2

The y-intercept is _____.

(0, -2)

9.5 Graphing Quadratic Equations in Two Variables

Complete the table and the graph.

x	y
-2	
-1	
	0
0	
1	
2	
	0
3	
4	

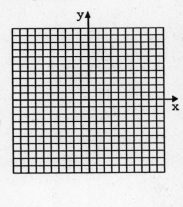

6
1
-.7
-2
-3
-2
2.7
1
6

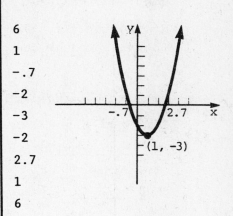

Graph each of the following. Identify the vertex.

9. $y = x^2 - 6x + 7$

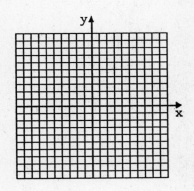

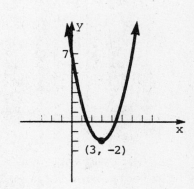

10. $y = x^2 + 4x + 1$

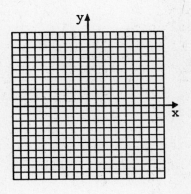

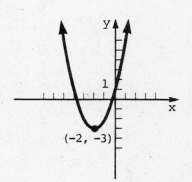

11. The graph of an equation $y = ax^2 + bx + c$ (*is/is not*) the graph of a function since for each value x (*is/is not*) exactly one value of y. This can be verified by the _____ _____ test.

    is

    is

    vertical line

12. The real number solutions of a quadratic equation $ax^2 + bx + c = 0$ are the x-values of the _____ of the graph of the corresponding function _____.

    x-intercepts

    $y = ax^2 + bx + c$

13. If the graph of the function $y = ax^2 + bx + c$ has two x-intercepts, then the corresponding equation _____ has _____ real solution(s).

    $ax^2 + bx + c = 0$; two

   If the graph of the function has one x-intercept, then the corresponding equation has _____ real solution(s). If the graph of the function has no x-intercept, then the corresponding equation has _____ real solution(s).

    one

    no

Decide from the graph how many real number solutions the corresponding equation has. Find any real solutions from the graph.

14. 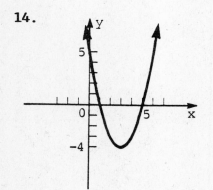 15.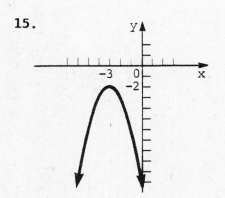

    two, 1 and 5; none

16. 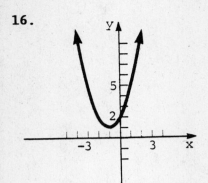 17. 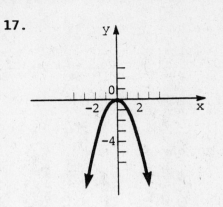 none; one, 0

Chapter 9 Test

The answers for these questions are at the back of this Study Guide.

Solve by using the square root property.

1. $m^2 = 32$

2. $(z + 4)^2 = 16$

3. $(2p - 5)^2 = 12$

Solve by completing the square.

4. $m^2 - 2m - 4 = 0$

5. $2z^2 + 5z = 12$

Solve by the quadratic formula.

6. $r^2 - 3r - 4 = 0$

7. $2p^2 - 5p - 1 = 0$

8. $\frac{1}{3}r^2 - r + \frac{2}{3} = 0$

Solve by the method of your choice.

9. $q^2 - 4q = 4$

10. $(3a + 2)^2 = 45$

11. $(3r - 1)(2r + 5) = 0$

12. $z^2 = 4z + 3$

1. _____

2. _____

3. _____

4. _____

5. _____

6. _____

7. _____

8. _____

9. _____

10. _____

11. _____

12. _____

13. $2x^2 - 7x = 4$

13. _____

14. $p^2 - 8p = 9$

14. _____

15. The manufacturer of a certain appliance has found that the monthly profit from the sale of x of these appliances is given by

$$p = 150x - x^2$$

where p is in dollars. How many appliances must be sold to produce a profit of $5625?

15. _____

Perform the indicated operations. Write the answers in standard form.

16. $(4 + i) + (2 - 3i) - (2 - i)$

16. _____

17. $(5 + 2i)(4 - 3i)$

17. _____

18. $(7 + 4i)(7 - 4i)$

18. _____

19. $\dfrac{2 - 3i}{3 + i}$

19. _____

Find the complex solutions of the following quadratic equations. Write the solutions in standard form.

20. $q^2 - 4q + 5 = 0$

20. _____

21. $2a^2 + 1 = 2a$

21. _____

Chapter 9 Quadratic Equations

Sketch the graph and identify the vertex of each parabola.

22. $y = -x^2$

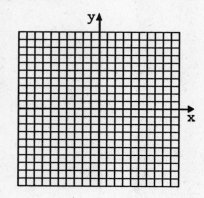

23. $y = x^2 + 4x - 1$

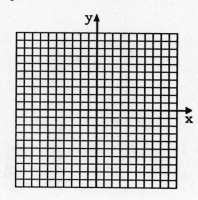

24. $y = -x^2 + 2x + 1$

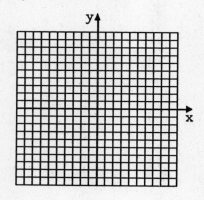

25. $y = x^2 - 4x + 2$

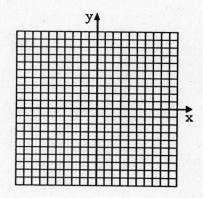

ANSWERS
TO
CHAPTER TESTS

ANSWERS TO CHAPTER TESTS

CHAPTER 1

1. 3/8 2. 31/18 3. 42/5 4. $-9 \geq -5$, false 5. $36 \leq 0$, false
6. $1/18 > -1$, true 7. $-9/8 \geq 0$, false 8. $12 \leq 36$, true 9. 6
10. $-37/7$ 11. $-|-3|$, or -3 12. .3709 13. $3x - 10$ 14. $\frac{4}{x+5}$
15. Yes 16. No 17. 16 18. 4 1/24 19. -5 20. 14
21. -28 22. $-3/2$ 23. -6 24. Undefined 25. -4 26. -7
27. -1 28. A, C 29. B 30. F, G 31. E, H 32. D, I
33. $-6z + 9$

CHAPTER 2

1. $8q$ 2. $-3p + 1$ 3. $-20z + 12$ 4. -1 5. -9 6. -9
7. 5/6 8. 0 9. $5 - 4x - 17; -3$ 10. $x + x + 9 = 71$; 31 men
11. $b = 2A/h$ 12. $b = P - a - c$ 13. True 14. 7/5 15. 8
16. $66.50 17. 15 ten-dollar bills 18. 3 hours
19. $15,000 at 12%, $23,000 at 14% 20. 150 liters 21. $q \leq -6$
22. $z \geq 0$ 23. $m > -3$ 24. $-1 \leq r < 9/5$ 25. At least 94

CHAPTER 3

1. 25/4 2. $-1/343$ 3. $1/9^2$ 4. 2^3 5. $\frac{1}{2^2 z^4}$ 6. $\frac{r^2}{5^3}$
7. 9.08×10^{-5} 8. 600 9. 200 10. degree 3; (a)
11. degree 2; (b) 12. $-13m^3 + 7m^2 - 16m + 8$ 13. $2m^3 - m^2 - 1$
14. $2r^2 - 12r + 5$ 15. $14p^5 - 28p^4 + 63p^3$ 16. $12q^2 + 13q - 14$
17. $12a^2 - ab - 35b^2$ 18. $10z^3 - 21z^2 - 29z + 28$ 19. $16p^2 + 72pq + 81q^2$
20. $9z^2 - 10zy + \frac{25y^2}{9}$ 21. $4m^4 - 49p^2$ 22. $-x - 2 + \frac{3}{2x}$
23. $\frac{3}{2} + \frac{2}{m} - \frac{1}{m^2} + \frac{1}{2m^3}$ 24. $2p^2 - 5p + 2$ 25. $5r^2 - 8r + 7 + \frac{-4}{3r^2 + 2}$

CHAPTER 4

1. $2xy(1 + 8x)$
2. $6ab(2a^2 - 3b^2)$
3. $3r^2(3 - 2r + 6r^2)$
4. $16r(2s + rs + 3r^2)$
5. $(p - 2)(p + 7)$
6. $(2 - m)(2 - q)$
7. $(p - 7)(p + 1)$
8. $5p^2q^3(p - 5)(p + 3)$
9. $(7m + 1)(m + 3)$
10. $(5r + 1)(2r - 7)$
11. $(3x - y)(2x + 3y)$
12. $(7z + 2y)(3z + 5y)$
13. $2(9p + 5)(9p - 5)$
14. $(5m + 7q)(5m - 7q)$
15. $(z^2 + 9)(z + 3)(z - 3)$
16. $(3p - 5)^2$
17. $(8z - 3)^2$
18. $2z(3z - 2)^2$
19. $(3z + 2)(9z^2 - 6z + 4)$
20. $(10p - 3q^2)(100p^2 + 30pq^2 + 9q^4)$
21. $(x + y)(x - y + 5)$
22. $-2/3, 5/2$
23. $2/3, -1/2$
24. $-2/3, 5/4, -3$
25. $0, 5, -5$
26. 8 inches
27. 3 cm
28. $-4 \leq r \leq 1$
29. $k < -3$ or $k > 1/2$
30. $z \leq -3/5$ or $z \geq 3/2$

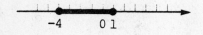
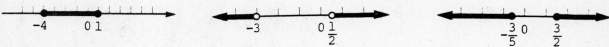

CHAPTER 5

1. $5, -3$
2. (a) $3/2$ (b) $-3/2$
3. $\dfrac{r^6}{2q^2}$
4. $\dfrac{r + 1}{2}$
5. $\dfrac{1}{pq}$
6. $-9/4$
7. $\dfrac{r + 2}{r + 6}$
8. $\dfrac{2p + 5}{2p - 5}$
9. $72y^3$
10. $(2m - 3)(m + 2)(3m - 4)$
11. $\dfrac{32z^3}{12z^4}$
12. $\dfrac{6m}{15m(m - 2)}$
13. $-\dfrac{3}{q}$
14. $\dfrac{-6}{5(m + 3)}$
15. $\dfrac{2a^2 - a - 2}{(a - 1)(a - 2)}$
16. $\dfrac{-2m^2 + 5m - 6}{(2m - 1)(m + 2)(2m - 3)}$
17. $\dfrac{1 + 5r}{1 + 3r}$
18. $\dfrac{r}{5 - 3r}$
19. 1
20. No solution
21. $4/5$
22. $2/3$ or 1
23. 10 miles per hour
24. $40/13$
25. 49

CHAPTER 6

1. (0, 3); (5, 0); (-5, 6); (10, -3) 2. (0, 0); (2, 6); (-1, -3); (-4, -12)

3. (9, 5); (9, 4); (9, 0); (9, -3)

4.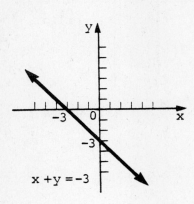

5.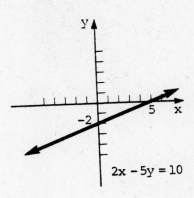

6.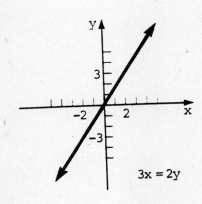

7.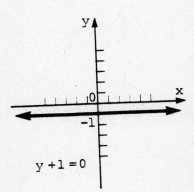

8. 4/3   9. -2/11   10. 0   11. 2x + y = -1   12. x + y = 4

13. 3x + 4y = -13

14.

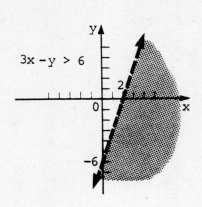

15.

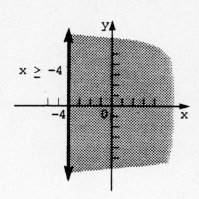

16. Function   **17.** Not a function   **18.** Function   **19.** Domain: set of all real numbers; range: $y \leq 0$ **20.** $-22; -6$

CHAPTER 7

1. (1, 4)   **2.** (3, 1)   **3.** No solution   **4.** (2, −1)   **5.** (19/3, −14/3)
6. (4, −3)   **7.** (−1, 7)   **8.** No solution   **9.** (−18, 16)
10. (8, −4)   **11.** (3, 6)   **12.** (1, 5)   **13.** (−6, 9)   **14.** 8, 17
15. 3 at $23 and 2 at $28   **16.** 40 liters of 70%, 80 liters of 40%
17. Boat: 20 mph, current: 5 mph

18.

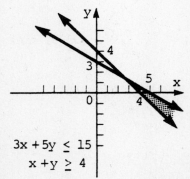

19.

20.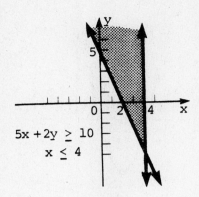

CHAPTER 8

1. 11   2. -60   3. -5   4. $4\sqrt{3}$ 5. $6\sqrt{3}/7$ 6. 32   7. $-2\sqrt[3]{2}$

8. $-\sqrt{5}$ 9. $10\sqrt{5} + 35\sqrt{3}$ 10. $2a\sqrt{2}$ 11. $38\sqrt{5p}$ 12. $25a^2b^5\sqrt{a}$

13. 79   14. $-5 + \sqrt{5}$ 15. $10 + 2\sqrt{21}$ 16. $2\sqrt{5}$ 17. $\frac{4z\sqrt{q}}{q}$

18. $\frac{\sqrt{14}}{4}$ 19. $\frac{7(1 - \sqrt{3})}{2}$ 20. 4   21. 1   22. 9   23. 8

24. 11^2 or 121   25. 5^2 or 25

CHAPTER 9

1. $4\sqrt{2}, -4\sqrt{2}$ 2. 0, -8   3. $\frac{5 + 2\sqrt{3}}{2}, \frac{5 - 2\sqrt{3}}{2}$ 4. $1 + \sqrt{5}, 1 - \sqrt{5}$

5. 3/2, -4   6. 4, -1   7. $\frac{5 + \sqrt{33}}{4}, \frac{5 - \sqrt{33}}{4}$ 8. 2, 1

9. $2 + 2\sqrt{2}, 2 - 2\sqrt{2}$ 10. $\frac{-2 + 3\sqrt{5}}{3}, \frac{-2 - 3\sqrt{5}}{3}$ 11. 1/3, -5/2

12. $2 + \sqrt{7}, 2 - \sqrt{7}$ 13. 4, -1/2   14. 9, -1   15. 75 appliances

16. $4 - i$ 17. $26 - 7i$ 18. 65   19. $\frac{3}{10} - \frac{11}{10}i$ 20. $2 + i, 2 - i$

21. $\frac{1}{2} + \frac{1}{2}i, \frac{1}{2} - \frac{1}{2}i$

22.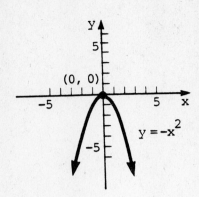

23.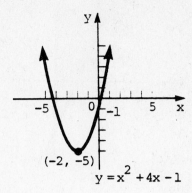

24.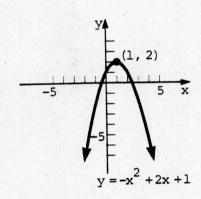

25.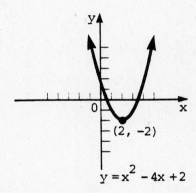

NOTES

NOTES

NOTES

NOTES

NOTES

NOTES